#47-3 6.00

POLLUTION CONTROL IN THE SOUTH AND NORTH

INDO-DUTCH STUDIES ON DEVELOPMENT ALTERNATIVES – 21

Indo–Dutch Studies on Development Alternatives

The present study is the twenty-first volume in the series sponsored under the *Indo–Dutch Programme on Alternatives in Development* (IDPAD). This programme grew out of intensifying contacts and cooperation in the seventies between Indian and Dutch social scientists interested in sharing each other's experiences in development research. Led by a common concern about the need to critically assess existing structures and policies from the point of view of meeting basic needs of large masses of people and to conceive alternatives that heighten the capabilities of the latter to serve as creative agents in the evolving world economy and society, a work plan of joint research was agreed on in 1980 forming the first phase of IDPAD (1981–83).

Since its inception, the programme has been jointly implemented by the Indian Council of Social Science Research (ICSSR), New Delhi, and the Institute of Social Science Research in Developing Countries (IMWOO), The Hague, and financed to a large extent by a grant from the Dutch government and a generous contribution from the Indian government. From January 1992 onwards, IMWOO's role as the Dutch counterpart organisation has first been taken over by Nuffic, the Netherlands Organisation for International Cooperation in Higher Education, The Hague, and from 1997 onwards, by WOTRO, the Netherlands Foundation for the Advancement of Tropical Research. Following the second (1984–89) and third phase (1990–96), the programme has entered its fourth phase (1997–2000) which has been based on the same principles as prepared by the Joint Committee of IDPAD.

The series of *Indo–Dutch Studies on Development Alternatives* includes only those research results obtained under IDPAD which are considered essentially qualified for wider dissemination. However, the views expressed and the facts stated therein are, of course, those of the authors.

A complete listing of the titles published in the series appears at the end of the volume.

POLLUTION CONTROL IN THE SOUTH AND NORTH

A Comparative Assessment of Environmental Policy Approaches in India and the Netherlands

O.J. Kuik
M.V. Nadkarni
F.H. Oosterhuis
G.S. Sastry
A.E. Akkerman

INDO-DUTCH STUDIES ON DEVELOPMENT ALTERNATIVES – 21

Sage Publications
New Delhi/Thousand Oaks/London

First published in 1997 by

Sage Publications India Pvt Ltd
M–32 Greater Kailash Market–I
New Delhi 110 048

Sage Publications Inc
2455 Teller Road
Thousand Oaks, California 91320

Sage Publications Ltd
6 Bonhill Street
London EC2A 4PU

Published by Tejeshwar Singh for Sage Publications India Pvt Ltd, phototypeset by Pagewell Photosetters, Pondicherry, and printed at Ram Printograph, Delhi.

Library of Congress Cataloging-in-Publication Data

Pollution control in the south and north: a comparative assessment of environmental policy approaches in India and the Netherlands / O.J. Kuik . . . [et al.].
p. cm.—(Indo-Dutch studies on development alternatives: 21)
(c : alk. paper)
Includes bibliographical references and index.
1. Pollution—Environmental aspects—India. 2. Pollution—Environmental aspects—Netherlands. 3. Environmental protection—India. 4. Environmental protection—Netherlands. 5. Environmental policy—India. 6. Environmental policy—Netherlands. I. Kuik, Onno, 1955– II. Series.
TD187.5.I4P65 1997 363.7'00954—dc21 97-27314

ISBN: 0–7619–9204–9 (US-HB)
81–7036–656–9 (India-HB)

Sage Production Team: *Evelyn George, O.P. Bhasin and Santosh Rawat*

CONTENTS

List of Tables 6
List of Figures 8
List of Abbreviations 9
Preface 11
Acknowledgements 14

PART I: BACKGROUNDS IN POLICY AND THEORY

1. Environmental and Economic Policy 17
2. Instrument Choice in Theory and Practice 30

PART II: COUNTRY STUDIES

3. Introduction 61
4. Policy Instruments for Pollution Control in India 67
5. Policy Instruments for Pollution Control in the Netherlands 116
6. Case Studies 168

PART III: CONCLUSIONS: COMPARISON AND POLICY IMPLICATIONS

7. Retrospective Summary 205
8. Similarities and Differences 224
9. Opportunities for Improvement in Pollution Control Policies 230

Annexure: A Comparison of Environmental Standards in India and the Netherlands 242
Bibliography 244
Index 250

LIST OF TABLES

2.1	Counter-Indications of Instruments	44
2.2	Government's Options to Influence the Behaviour of Economic Agents with respect to the Environment	51
3.1	Some Basic Indicators of Economy—1981 and 1991: Karnataka, India, and the Netherlands	62
4.1	Categorisation of Industries and Frequency of Inspection	84
4.2	Water Cess Levied on Water Consumption by Type of Use	89
4.3	Water Cess Paid by Industrial Units in India	90
4.4	Proportion (%) of Industrial Units with Pollution Treatment Plants in Operation in Karnataka	97
4.5	Prosecutions Launched and Convictions Secured under Pollution Control Laws	99
4.6	Trends in Water Quality of Rivers at Selected Locations in Karnataka State	99
4.7	Sources of Finance of Karnataka State Pollution Control Board (KSPCB)	101
4.8	Cost of Enforcement Incurred by KSPCB—Category-wise	106
4.9	Annualised Costs of Pollution Control Relative to Turnover of Selected Industries	107
4.10	Cost of Compliance of Pollution Control per Factory in Cement, Textile, and Fertiliser Industries	109
4.11	Percentage Share of Highly Polluting and Other Industries in All Industries for Important Indicators: 1980–81 and 1988–89—Karnataka and India	111
4.12	Growth of Highly Polluting Industries vis-à-vis Other Industries	112
5.1	Development of Waste Supply and NEPP-plus Target	153
5.2	Costs of Dutch Environmental Policy per Theme	158
5.3	Indicators for the Enforcement of Environmental Law	159
5.4	Costs of Industry's Own Environmental Activities	161
5.5	Environmental Costs in Dutch Industry	162
5.6	Actual and Projected Environmental Costs for Industry	163
6.1	Textile, Cement and Fertiliser Industries in Relation to the Total Industrial Sector	169
6.2	Pollutants vis-à-vis Standards in the Case of a Textile Mill	176
6.3	Cost Burden on Account of Pollution Control in Textile Industry	177
6.4	Status of Pollution Control in Cement Industry	185
6.5	Cost Burden on Account of Pollution Control in Cement Industry	185

6.6a Production, Consumption and Imports of Chemical Fertilisers in India **187**
6.6b Production, Consumption and Imports of Chemical Fertilisers in the Netherlands **187**
6.7 Standards Prescribed by KSPCB for Fertiliser Industry **189**
6.8 Status of Pollution Control in Fertiliser Industry in India (1991) **190**
6.9 Emissions from Dutch Fertiliser Industry, 1988 **191**
6.10 Situation–Characteristic Indicators and Policy Instruments of the Bovenberg-Framework **195**
6.11 Situation–Characteristics and Policy Instruments for Bio-degradable Waste Water Emissions from the Textile Industry **199**
7.1 Summary of Case Studies of the Textile, Cement and Fertiliser Industries in the Netherlands and Karnataka **222**

LIST OF FIGURES

4.1	The Pollution Abatement Policy Enforcement Mechanism	86
5.1	Composition of Charge Revenues	139
5.2	Development of the Average Charge Level and Waste Water Emissions by Firms	141
5.3	Environmental Indices, Targets and GNP Growth in the Netherlands	145
7.1	Institutional Structure of Pollution Control in India	214
9.1	Impact of Production Growth and Technological Development on Environmental Costs	238

LIST OF ABBREVIATIONS

Aeq	Acid equivalents
AER	Algemene Energie Raad (General Energy Council)
alara	as low as reasonably achievable
ASI	Annual Survey of Industries
BAT(neec)	Best Available Technology (not entailing excessive cost)
BOD	Biochemical Oxygen Demand
BWSSB	Bangalore Water Supply and Sewerage Board
CAC	Command and Control
CBS	Central Bureau of Statistics
CFC	Chlorofluorocarbons
CO	Carbon monoxide
CO_2	Carbon dioxide
COD	Chemical Oxygen Demand
CPB	Central Planning Bureau
CPCB	Central Pollution Control Board
CRMH	Central Council for Environmental Protection (Centrale Raad voor de Milieuhygiëne, since 1993 'Raad voor het Milieubeheer')
DAP	Di-Ammonium Phosphate
Dfl.	Dutch guilder
DGM	General Directorate Environmental Management, VROM
DOE	Department of Environment
EC	(Commission of the) European Communities
EIA	Environmental Impact Assessment
ESP	Electrostatic Precipitator
ETP	Effluent Treatment Plant
EU	European Union (former European Community)
GATT	General Agreement on Tariffs and Trade
GDP	gross domestic product
GNP	gross national product
IFCI	Industrial Finance Corporation of India
kl	Kilolitre
KSPCB	Karnataka State Pollution Control Board
KWS 2000	'Hydrocarbons 2000' action programmes
MAR	Maximum Acceptable Risk
MEF	Ministry of Environment and Forests
MIC	Methyl Iso Cyanate
MINAS	Minimum National Standard

NAFTA	North American Free Trade Association
NCEP	National Committee on Environment Planning
NCEPC	National Committee on Environmental Planning and Coordination
NEPP	National Environmental Policy Plan
NEPP-2	Second National Environmental Policy Plan
NGO	Non-governmental Organisation
NH_3	Ammonia
NO_x	Nitrogen oxide
N_2O	Dinitrogen oxide
NOC	No Objection Certificate
OECD	Organisation for Economic Cooperation and Development
PCB	[1] Pollution Control Board [2] Polychlorobifenyl
PIL	Public Interest Litigation
PPP	Polluter Pays Principle
PU	Pollution Unit
PVC	Polyvinylchloride
RARO	Raad van Advies voor de Ruimtelijke Ordening (Advisory Council for Town and Country Planning)
R&D	Research and Development
RIVM	National Institute of Public Health and Environmental Protection
RMNO	Raad voor het Milieu- en Natuur Onderzoek (Council for Environmental and Nature Research)
Rs	Rupees
SBI	Standaard Bedrijfs Indeling (Statistical Classification of Industries in the Netherlands)
SER	Sociaal-Economische Raad (Socio-Economic Council)
SO_2	Sulphur dioxide
SPCB	State Pollution Control Board
SPM	Suspended Particulate Matter
SS	Suspended Solids
SSIs	Small Scale Industries
SWPC	Surface Water Pollution Charge
TDS	Total Dissolved Solids
TSP	Total Suspended Particulates
TSS	Total Suspended Solids
US	United States
VOCs	Volatile Organic Compounds
VROM	Ministry of Housing, Physical Planning and Environment
Wabm	General Environmental Provisions Act (1980)
WCED	World Commission on Environment and Development
WQM	Water Quality Manager
WRR	Wetenschappelijke Raad voor het Regeringsbeleid (Scientific Council for Government Policy)

PREFACE

Environmental pollution is a common concern of almost all countries in the world, whether rich or poor. A path of economic growth that causes unchecked damage to the environment is neither desired nor considered to be sustainable by most governments. Thus, most countries in an effort to address the problems have set up some form of environmental policy, attuned to their specific circumstances and needs.

Environmental policy is a relatively 'new' government activity and therefore in the midst of finding its best manifestation in an increasingly complex world. Environmental policy makers are continuously learning from their past mistakes and successes. The challenge is to find that mix of policy instruments which is simultaneously effective, efficient, equitable, administratively feasible, and acceptable.

India and the Netherlands have a relatively long history in environmental policy. Their starting points, however, differed fundamentally. India is a developing country in the process of modernisation while the Netherlands has a highly developed economy entering into what is sometimes called the post-industrial era. Can both countries learn from each other's experiences in environmental policy approaches despite these differences? The possibility of an answer in the affirmative inspired the undertaking of the present study 'A Comparison of Policy Instruments for Pollution Abatement in India and the Netherlands' in the framework of the Indo-Dutch Programme on Alternatives in Development (IDPAD). The study has been jointly carried out by researchers at the Institute for Social and Economic Change (ISEC, Bangalore, India) and the Institute for Environmental Studies at the Vrije Universiteit (IES, Amsterdam, the Netherlands). The main objectives of the study are:

1. To provide insight into the constraints and strengths of pollution control instruments in India and the Netherlands.
2. To recommend improvements in the efficiency and effectiveness of policy instruments aimed at abating and preventing pollution in India and the Netherlands.

The study emphasises, but is by no means limited to, the role of 'economic' instruments in abating pollution of air and water by industrial activities. Economic instruments deserve special attention because they offer, at least in theory, considerable potential for improvements in the efficiency of environmental policies. The emphasis on air and water pollution by industries is not meant to suggest that other environmental problems are unimportant, but we believe that these unwanted side effects of rapid industrial growth justify the mentioned focus. Geographically, the study focuses on the Netherlands and the state of Karnataka in India.

The book consists of three parts: (*a*) a general introduction to the theory and practice of environmental policy, (*b*) a detailed description of environmental problems and policy approaches in both countries, and (*c*) the comparison and synthesis of conclusions in terms of policy recommendations.

Part I of the book describes the research area from an international perspective. Chapter 1 focuses on the main trends in environmental policy making, with emphasis on the potential and actual performance and acceptance of economic instruments. Chapter 2 presents a theoretical framework to assess the performance of different types of policy instruments in specific policy environments. Part I ends with the formulation of the main hypotheses and research questions of the current study.

Part II presents country case studies. For both India (Chapter 4) and the Netherlands (Chapter 5), a description is given of the environmental policy instruments used, their historical development and the main environmental problems they are supposed to tackle. An assessment of these instruments is carried out, both in terms of environmental effectiveness and in terms of administrative and financial burden, including the distribution of this burden. Attention is also paid to the practical feasibility, institutional compatibility and concordance with prevailing policy principles. Chapter 6 presents in-depth case studies of policy approaches towards three branches of industry in India and the Netherlands, namely textiles, cement and fertiliser.

Part III brings together the results of Parts I and II. Similarities and differences in both the performance of policy instruments and the stated factors regarding the policy environment are identified and explained. Lessons are learnt from each other's experience and the potential for improvements is explored.

We would like to thank IDPAD for making this study possible. We also gratefully acknowledge the help of the Department of Spatial and Environmental Economics of the Vrije Universiteit of Amsterdam. Ms M. van Amelrooy, a student at the department, wrote a very useful thesis on the use of economic instruments in Indian environmental policy. Prof. J.B. Opschoor and Prof. H. Verbruggen encouraged us all the way and helped us by detailed comments on earlier drafts. Valuable contributions to the study have also been made by Ms K. van den Berg and Mr J. Nugteren. Comments by two anonymous referees (one from India and another from the Netherlands) have helped to improve the manuscript. Finally, thanks are due to the many officials, scientists and industrialists who kindly provided us with information and with whom we discussed policy matters. Any errors in the text remain, of course, the responsibility of the authors.

Bangalore/Amsterdam
1996

O.J. KUIK
M.V. NADKARNI
F.H. OOSTERHUIS
G.S. SASTRY
A.E. AKKERMAN

ACKNOWLEDGEMENTS

The project on which the present publication is based was funded by the Indo-Dutch Programme on Alternatives in Development (IDPAD) in collaboration with the Indian Council of Social Science Research (ICSSR), New Delhi, and the Institute for Social Science Research in Developing Countries (IMWOO), The Hague. IDPAD, however, does not take responsibility for the facts stated, opinions expressed and conclusions reached in this publication.

Part I

BACKGROUNDS IN POLICY AND THEORY

1

ENVIRONMENTAL AND ECONOMIC POLICY

This chapter is devoted to the relationship between environmental and economic policy. Section 1.1 looks at the 'tool box' of environmental policy instruments. Section 1.2 discusses the increased interest for putting into practice the old idea of applying economic instruments to environmental problems. Sections 1.3 and 1.4 deal with some recent trends in the main features of environmental policy in developed and developing countries, respectively. Finally, Section 1.5 concludes with some considerations regarding the possible antagonisms and synergism between economic and environmental policy in both industrialising and industrialised countries.

1.1 INSTRUMENTS OF ENVIRONMENTAL POLICY

The success of any environmental policy depends on changes in the behaviour of producers and consumers. The environmental authorities can try to bring about these changes by means of various instruments. Unfortunately, there is no generally accepted, standardised and unequivocal classification of environmental policy instruments. However, there is some agreement that three broad categories of instruments can be distinguished (see Opschoor & Turner 1994, p. 10):

1. Instruments, aimed at voluntary adaptations of individual and group behaviour in a more environment-friendly direction. This category, to which we will refer as 'communicative instruments', includes among others:

- the provision of knowledge and information in all possible forms, both on the environmental effects of the present behaviour and on 'cleaner' alternatives;
- moral suasion, i.e., campaigns to persuade people and institutions to change their behaviour;
- institutionalisation of environmental management within firms and other organisations;
- voluntary commitments by trade and industry, or agreements between them and the government, implying certain environmental targets to be realised (according to a specified time schedule).

2. Instruments which affect the market conditions under which people and firms make their decisions, without directly reducing the decision space available to them. This category is commonly referred to as 'economic instruments'. Within this category, a distinction can be made between:

   - charges and taxes, which make polluting behaviour more expensive;
   - subsidies and other types of financial support (such as tax reductions), which make environment-friendly behaviour relatively cheaper;
   - tradable rights(to emit a certain amount of pollutants,or to produce or use a certain amount of polluting goods or substances);
   - deposit–refund systems, in which a deposit for a potential polluting product is being paid by the purchaser, who can claim a refund after returning the product or showing that the pollution did not take place. These systems have the effect that a product (or substance) keeps a certain monetary value, even if it has become useless to the one who bought it. In this way, the uncontrolled disposal of specific waste streams can be prevented;
   - public procurement, by which the government an set a good example and may pave the way for cleaner products and technologies;
   - liability legislation, requiring the polluter to compensate the environmental damage he caused, and thus providing a financial incentive for pollution prevention (either directly, or indirectly, through higher insurance premiums).

3. Instruments which influence the range of alternatives by means of prohibitions, restrictions or obligations ('direct regulation'). This can be done by introducing product, process or emission standards, by prescribing how certain activities have to be performed, demanding certain professional skills, specifying the precautionary measures to be taken, etc. Outright bans on certain activities, products or substances also belong to this category. Bans may be limited in space and time: they can, for instance, relate only to vulnerable areas or smog-promoting weather conditions. Finally, the provision of infrastructure and facilities (dustbins, cycle tracks) may broaden the range of available alternatives by including environment-friendly ones, instead of limiting it by excluding environmentally harmful ones. This type of public investment can also be regarded as a form of direct regulation.

At first glance, the tool box of the environmental policy maker thus seems to be filled with a large variety of instruments. In practice, the scope for choosing between alternative instruments is limited by various factors. An instrument should fit in with the legal and institutional[1] framework (although one could try to change this framework if there is a strong preference for a specific instrument) and it should be technically and administratively feasible. The information, needed to apply the instrument, has to be available. There should be enough support, or at least not too much opposition against the instrument in society, to prevent sabotage and to keep enforcement costs reasonably low. The costs and other economic consequences of applying the instrument should be reasonable and fairly spread among groups in society. Especially in developing countries, adverse effects on the lowest income groups should be avoided as much as possible. Basic principles of environmental policy, such as 'the polluter pays', 'prevention is better than cure', the 'precautionary principle' and the 'stand-still principle' should be met (if they are accepted by the policy makers, of course). The instrument should have no significant negative side effects, either within the environmental sphere (for instance, by transforming air pollution into water or soil pollution) or in terms of other social and political objectives (e.g., employment, distribution of income).

Within these limitations (which are sometimes rather vague) one can try to find the most effective, or the most efficient instrument; in other words, an instrument which maximises the effect[2] (given a certain budget) or which minimises costs (given a certain environmental objective). In doing so, the dynamic features of society should be kept in mind. If economic activity is increasing, the increase may often more than off-set a possible decrease of pollution per unit of output or consumption. Policy instruments which may be effective in a stable situation, may be less so in this dynamic context.

Another dynamic aspect of policy instruments is their influence on the technological development. Some instruments contain a continuous incentive to innovate in cleaner techniques; others do not. This feature should be considered when assessing the effectiveness of an instrument.

1.2 BACKGROUNDS OF THE QUEST FOR COST EFFECTIVE INSTRUMENTS

The publication of the report 'Our Common Future' by the World Commission on Environment and Development (in 1987) may be regarded as a benchmark in understanding the relationship between environmental and economic policies. Among the critical objectives for environment and development policies that follow from the concept of sustainable development the report includes 'merging environment and economics in decision making' (WCED 1987, p. 49). This requires changes in the legal and institutional frameworks that will enforce the common interest. Moreover, environmental objectives must be built into taxation, prior approval procedures for investment and technology choice, foreign trade incentives, and all components of development policy. When discussing strategies for sustainable industrial development, the WCED advocates the use of economic instruments (WCED 1987).

The United Nations Conference on Environment and Development (UNCED), which took place in Rio de Janeiro in 1992, has confirmed the need for such an 'economic' approach to environmental policy making (and vice versa). At various places in the main outputs of the conference, there is recognition of the 'polluter (and user) pays principle', the need to internalise environmental costs, the precautionary approach to environmental change, and the use of economic instruments ('Declaration of Rio' and 'Agenda 21').

The search for more efficient tools for environmental policy is also going on at the OECD level. A recent OECD report mentions several advantages of economic instruments (OECD 1991, p. 14):

- They can yield substantial cost-savings by allowing polluters to determine the most appropriate ways of meeting a given standard (as in the case of tradable rights), or by equating the marginal cost of environmental protection to the level of emission charges across the whole range of activities.
- They offer an ongoing incentive to reduce pollution below the levels determined by regulations. They also encourage new pollution control technology, production processes and new non-polluting products by favouring new research and development activities.
- They increase flexibility. For the authorities, it is often easier and faster to modify and adjust a charge than to change legislation or regulation; for polluters, the freedom to choose within an overall financial constraint is preserved.
- They can promote resource conservation and transmission to future generations in the same way that resource-pricing does.
- They may provide a source of finance, which may either be directed to specific environmental programmes, or used as a significant instrument in overall fiscal policy.

The OECD guidelines for applying economic instruments pertain specifically to environmental charges or taxes, marketable permits and deposit–refund systems. Subsidies, which are also an often used type of economic instrument, are generally incompatible with the Polluter Pays Principle (PPP) which the OECD Council adopted in 1972.

The PPP is also one of the guiding principles for the environmental policy of the European Union. Other principles are:

- the prevention principle, which states that pollution should preferably be prevented rather than abated after its emergence;

- the precautionary principle, which explicitly recognises the existence of uncertainty (environmental and social) and seeks to avoid irreversible damages via the imposition of a safety margin into policy;
- the economic efficiency/cost effectiveness principle, applying both to the setting of standards and the design of the policy instruments for attaining them;
- the subsidiarity principle, to assign environmental decisions and enforcement to the lowest level of government capable of handling it without significant residual externalities;
- the legal efficiency principle, to preclude the passage of regulations that cannot be realistically enforced.

To economists, the idea of combating pollution efficiently by internalising external costs is nothing new. Pigou already advocated this as early as 1920. From 1960 onwards, and especially since 1975, a vast literature on the theoretical merits and demerits of economic instruments has emerged (for a review, see Bohm & Russell 1985). Reviews of practical applications were published by the OECD (Opschoor & Vos 1989; Opschoor et al. 1994). By the end of the 1980s, it had become clear that much of the original debate on economic instruments had remained too remote from the realities facing environmental decision makers. A more pragmatic (but still cost effective) approach was advocated by economists.[3]

To summarise, at the beginning of the 1990s, there seems to have grown a broad and firm support for economic and other cost effective instruments for environmental policy, at least at the level of principles, ideas and theories. To put all this into practice, however, requires a fertile breeding ground at the level of actual policy making. In the following sections, we will have a brief look at the environmental policy framework in both the industrial and the developing world.

1.3 TRENDS IN ENVIRONMENTAL POLICY IN DEVELOPED MARKET ECONOMIES

Until recently, western market economies have generally relied almost entirely on 'command and control' types of environmental policy. This kind of policy has shown some successes, notably in the area of air and water pollution, but the limitations of this approach are becoming evermore visible. They can be summarised as follows:

- In many cases, environmental problems have been shifted instead of solved. A well-known example is the sludge from water purification plants which is often so heavily contaminated by substances like heavy metals that there is no productive use for it.
- The speed of emission reduction is too low to reach the environmental objectives within a reasonable time. To a large extent, these disappointing results can be attributed to production growth, undoing a large part of the effects. Lengthy and detailed legislation and licensing procedures are also contributing to delays in implementation. Furthermore, new environmental problems have emerged which had not been recognised before, making even more stringent requirements necessary.
- A vast amount of non-point and mobile sources (consumers, cars, small-scale manufacturing, agriculture) has replaced a limited number of point-sources (large-scale industry, power plants, etc.) as the major contributor to many kinds of pollution. This creates increasing monitoring and enforcement problems.
- The traditional way of dealing with polluters, i.e., by prescribing measures or standards for each individual source, is felt to be inefficient and inflexible. On the one hand, there is often no incentive for existing sources to innovate in cleaner products and processes. On the other hand, new sources often find themselves in an unfavourable competitive position in comparison with existing sources because they have to comply with tighter standards. This is also an obstacle to innovation.

At the same time, it is being recognised that the so-called 'target groups' of environmental policy (e.g., consumers, branch organisations, public utilities, etc.) could be involved more actively in its implementation. Together with the 're-discovery' of the market as a regulating mechanism, this has created a favourable atmosphere for a policy style which leaves more responsibility and initiative to the polluters themselves, and which stimulates imaginative solutions. This calls for instruments using the market mechanism and appealing at both the self-interest and the environment-consciousness of the individual polluter. Charges, tradable emission or production rights, liability, voluntary agreements and instruments of communication contain these features to a larger extent than 'command-and-control' instruments.

Another major trend in western environmental policy is towards various kinds of integration:

- integration across the borders of separate parts of the environment (water, air, soil);
- integration of environmental policy with other areas of government policy which may affect the environment;
- integration over the entire life cycle of substances and products (the so-called 'cradle-to-grave' approach, or 'integrated chain management');
- integration across national borders (especially in those areas where the pollution has transboundary effects: global warming, stratospheric ozone depletion, acidification, marine and river pollution).

The degree to which these types of integration have been implemented differs from country to country. The third type (integrated chain management) especially is still in its infancy. Anyhow, all kinds of integration point in the direction of a more coordinated way of applying environmental policy instruments. There is a growing recognition that isolated measures regarding single pollutants, single products and single parts of the environment are no longer appropriate as the core of environmental policy. Instruments should have a more universal applicability than at present. This might well lead to a future environmental policy stating ambient environmental quality targets at a policy level which may vary with the issue at hand (from local to global). The fulfilment of this target could then be a collective responsibility of all polluters concerned, and could be realised by tradable rights and/or voluntary agreements, supported by a system of taxes and charges on basic inputs (like non-renewable energy and raw materials) which shifts the burden of taxation from labour and capital to natural resources.

The collapse of the centrally planned economies of the former Soviet bloc has reinforced an upsurge of sympathy for, and interest in, market based approaches and market analogues to be used across a range of policy contexts. The phrase 'market-oriented environmental policy' itself may now be assuming some political value (Stavins & Whitehead 1992). Confidence in all kinds of government regulations is understandably low in these countries. The economies in transition appear to offer good tested conditions for economic instruments as they have to revise large parts of their institutional and fiscal systems.

1.4 TRENDS IN ENVIRONMENTAL POLICY IN DEVELOPING COUNTRIES

The time that environmental problems were seen as a luxury problem, with which developing countries had nothing to do, has gone now. The Brundtland Commission's report 'Our Common Future' (WCED 1987) can be regarded as a milestone in this respect. It is now widely being recognised that poverty and environmental degradation are often mutually reinforcing (which is not the same as saying that economic growth automatically provides the solution to pollution).

The rural environment in developing countries is being threatened by deforestation, loss of fertile land (by soil erosion, salinisation and overgrazing), depletion and degradation of water resources, and loss of biodiversity. Large-scale projects and operations, such as mining and hydropower plants are also taking their toll in the countryside. Overfishing and water pollution have wiped out fish stocks in many rivers, lakes and coastal waters. In urban areas, air, water and soil are being polluted by emissions from households, traffic, industry, power plants and waste dumps. The combination of proximity between industrial and residential areas and a lack of safety measures creates a permanent risk of disaster for many people.

Though problems of land degradation and deforestation are considered grave issues in developing countries, accelerated industrialisation in some of these countries like India has increased the importance of tackling industrial pollution. This is not just for aesthetic and health reasons alone as in developed countries but also because industrial pollution is threatening livelihoods. For example, marine pollution has been affecting fisheries; fly ash dumps are at the expense of agricultural lands; and cement dust is reducing crop yields in neighbourhood farms.

The response of governments in developing countries to the environmental challenge has been mixed, but generally speaking insufficient to turn the tide. It is only fair to say that the means to do so are far more limited than in the developed countries. Poor people often have no other choice if they want to survive than to overcharge the environment. Widespread bureaucracy and corruption do not favour efficient environmental policy implementation. Finally, the debt burden of many countries leaves no room for additional public spending.

In conclusion, environmental problems in developing countries are not entirely the same as those in developed market economies, where they are often a side effect of material affluence. The feasibility of applying particular environmental policies also differs between countries. Therefore, copying each other's environmental policy system would be inadequate, even if this system would be perfect for one country.

Nevertheless, the legislative and institutional framework for environmental policy in developing countries has in many cases been developed along the same lines as in the western market economies. This is especially true for industrial emissions. The emphasis is on combating single pollutants, mainly by a 'command and control' approach. However, a change towards more integration is already visible, for example, in the Indian Environment Protection Act of 1986. Some of the main principles of western-type environmental policy have not yet penetrated the South (as a matter of fact, they are also often not applied in practice in the North). Investment subsidies, depreciation allowances and reductions in tax and tariff rates are very common, although they contravene the PPP. The 'precautionary principle' is also scarcely being applied: very often, measures are only being taken after severe damage has already occurred. A large part of environmental policy measures is curative instead of preventive.

The role of NGOs differs from those in the North, where environmental organisations are often recognised as a discussion partner by the government and try to influence policy. In the South, their role is more to organise people on specific issues, for instance, mobilising support to get a polluting plant closed down. Sometimes, concerned citizens start lawsuits against polluters or against the state, as has been done for example in India by the active environmentalist Mr M.C. Mehta.[4] These lawsuits may demand pollution abatement measures, compensation for damage, or closure of the polluting firm. The role of these private actions in developing countries is undoubtedly very important.

1.5 ENVIRONMENTAL AND ECONOMIC POLICY: HARMONY OR CONFLICT?

Developing countries, especially those receiving financial support from IMF and World Bank, have undertaken major revisions of their economic policy recently. Price subsidies are being curtailed, government spending in general is being reduced, capital markets are being liberalised. Exchange rates have been adjusted and quite a number of import restrictions have been lifted. The role of government regulation in general has become more modest. Large parts of the economy, once sheltered by state ownership, monopoly, and/or protectionism, now face the merciless competition from and on the world market.

In India, the first steps towards liberalisation of foreign trade have already been taken in the late 1970s. This has led to a substantial increase in both imports and exports during the 1980s. The net result, however, was an increasing deficit on the balance of payments. The government budget deficit has also grown, both as a result of increases in public spending and subsidies, and of lagging tax income. In 1991, the Finance Minister, Manmohan Singh, introduced a package of measures to cure the Indian economic troubles. These 'New Economic Policies' (NEP) entail, among other things, fiscal reforms, a decrease of subsidies (e.g., those on fertilisers), and price increases of oil, public transports and other products of public enterprises. Exports are being stimulated by tax exemptions for exporters and devaluations of the rupee (Storm & Naastepad 1991).

The consequences of the NEP and other Structural Adjustment Programmes (SAPs) for the environment are heavily debated. Not surprisingly, the advocates of liberalisation argue that the net effects are positive and point to the fact that the new prices will reflect scarcities better, and will therefore lead to a more efficient allocation and a more rational use of scarce resources.[5] Certain public spending programmes which may be harmful to the environment, such as road construction and irrigation schemes, are being cut down. Property rights may be better defined and allocated, and this could lead, for instance, to a better management and conservation of agricultural land.

Opponents, on the other hand, argue that the new emphasis on cash crops and export-oriented industry will lead to further environmental degradation and pollution, accompanied by increasing income disparities and a more vulnerable agriculture (Kothari & Kothari 1992). They fear that opening up the economy will attract foreign investors who show no respect to the environment, and cite with horror World Bank economist L. Summers, who stated that certain parts of the Third World are still 'underpolluted'. The public expenditure cuts they mention are those which hit the environment-oriented items: pollution control programmes, water supply and sanitation investments, alternative energy promotion, etc. Sometimes, other negative developments are being ascribed to the SAPs. Kothari & Kothari (1992) point to the fact that as a result of the conversion to cash crops, large parts of the rural population may have to resort to livelihoods other than agriculture. They will settle in the rainforest or move to the cities, thereby aggravating the already severe environmental problems.

Similar discussions can be observed in the northern world. Especially the liberalisation of international trade (EU, NAFTA, GATT) has recently been surrounded by fierce discussions on the advantages and disadvantages of free trade for the environment. Those who argue that a borderless market leads to a more efficient use of scarce resources (and thus leads to environmental benefits) find themselves confronted with those who predict environmental disaster because of the growth in transport and relocation of 'dirty' industries in the poorest countries.

In a sense, both points of view are justified, but one might also say that both miss the point: SAPs and trade liberalisation have no environmental, but economic objectives. They cannot be expected to lead to improvements in environmental quality automatically. They will, however, have environmental impacts, some of them positive, others (probably more, at least in the short run) negative. It is essential that a country which embarks on a SAP and/or opens it borders keeps the power to mitigate the negative effects by introducing and/or maintaining an adequate environmental policy. Such a policy will often imply more government intervention and higher production costs. It could also lead to increases in public spending (although this can be limited by applying the PPP). This tendency seems to be at odds with the wish to compete in the world market, to cut back bureaucracy, and to clean up public finance. However, effective and efficient environmental policies are needed to approach both the welfare–economic ideal of optimal allocation and the economic–ecological ideal of sustainable development. In order to make such policies compatible with structural adjustment and liberalisation, a careful choice of targets and instruments is required. Environmental policy makers should be aware of the choices available to them, and take advantage of this scope in order to minimise the negative effects and maximise the positive effects for both the environment and the economy. Although economic instruments are not a panacea, they should at least be part of the considerations and might in certain cases provide the 'double dividend' of economic reform and environmental improvement better than other instruments.

Notes

1. This framework also includes international treaties (e.g., the Montreal Protocol), and the regulations issued by higher authorities (e.g., the federal government in India or the European Union for the Netherlands).
2. Including possible positive side effects.
3. See, for instance, Baumol & Oates (1990); Hahn (1989).
4. See Bhat (1992, p. 41); Bowonder & Arvind (1989).
5. Jansen (1991) rightfully comments that this does not mean that these prices include a payment for external (environmental) effects. The 'real' scarcity prices would be higher than the free market prices.

2

INSTRUMENT CHOICE IN THEORY AND PRACTICE

Command and control (hereafter CAC) type instruments are common in current environmental policy in the North and South (see Sections 1.3 and 1.4). It is increasingly realised that the demands for a broadening and intensification of environmental policy cannot be met with this type of instrument alone. Shortcomings of the CAC type instruments that are often mentioned are: non-optimal allocation of abatement measures across sources and large information and enforcement costs to the administrator. It is a well-established result of economic theory that if polluting sources face different marginal abatement costs, a CAC ('standard-setting') approach to pollution control will be more expensive than market based approaches which allow these sources more flexibility in controlling pollution. An important proposition in the economics of pollution control is that the cost of achieving a given reduction in emissions will be minimised if the marginal costs of control are equalised for all emitters (Tietenberg 1992a). In most situations CAC type instruments will not produce this least-cost solution. In the United States, a number of studies have compared the costs of the current regulatory (CAC) approach of US air pollution control with the (calculated) least-cost ideal. The ratio of CAC costs to the least-cost ideal ranges from 1.78 to 22, that is, the current control costs are 78 to 2100 per cent more expensive than theoretically necessary to meet the ambient air quality standards required by US law (Tietenberg 1992a). These cost differences are impressive; they warrant the search for more efficient instruments for pollution control.

A second drawback of the traditional CAC instruments of environmental policy is their increasing burden on the administrator. Exponentially rising information and enforcement costs of expanding and increasingly complex environmental regulations are bound to sooner or later hit the limits of the capabilities and means of any administration.

Efficiency and administrative feasibility are the main arguments for the search of new (combinations of) environmental policy instruments. Since long, economists have advocated the use of market based regulatory instruments (charges, tradable permits).[1] More recently attention is also focused on the complementary role of private, judicial instruments (see Tietenberg 1992b).

As more policy instruments are identified and added to the instrument set which *could* be used, the question that naturally comes up is which instruments *should* be used for particular purposes and in particular circumstances. Policy-oriented theories of optimal instrument choice have been developed in recent years. Section 2.2 briefly reviews some of the literature on this subject. However, before that, the following section examines the criteria to judge the performance of policy instruments in the environmental field.

2.1 CRITERIA

In order to judge the performance of policy instruments, performance criteria should be established. Several criteria have been suggested by different authors; it would be of little use to reiterate all these criteria here. In a recent publication, OECD has distinguished five sets of criteria to assess the performance of environmental policy instruments (OECD 1991).

Environmental effectiveness

The main purpose of environmental policy instruments is to achieve certain environmental goals or targets. One can say that environmental effectiveness is the single most important criterion. An instrument which is not effective in reaching its target does not perform well, irrespective of its performance with respect to other criteria. Effectiveness depends on the compliance of the polluting agents. As discussed in Section 1.1, this compliance can be of a voluntary nature (communicative instruments), there can be financial incentives to comply (economic instruments), or it can be legally enforced (direct regulation). Effectiveness is essentially a dynamic concept. Environmental policy instruments are applied in a changing world. A set of measures which has been effective in one set of circumstances (economic and environmental) is not necessarily effective in another set. Instruments should therefore be able to adjust to changing economic and environmental circumstances.

Economic efficiency

Environmental policy instruments should attain environmental targets at least (social) costs. Inefficient approaches to pollution control waste scarce resources which could be used in other ways to generate higher levels of environmental quality or to provide other goods and services. With respect to efficiency a distinction should be made between static and dynamic, or short-term and long-term efficiency. Instruments which offer the same static or short-term efficiency may differ in their dynamic efficiency, namely, the incentive they offer to search for new, environment-saving technology in the longer run. Another distinction that can be made is between instruments that distort relative prices (by for instance making capital equipment artificially cheap) and those that do not.[2]

Equity

As a rule, environmental policy instruments affect the distribution of income and wealth. In a way, they are meant to do so. Traditional rights (to pollute or to use-up resources) are taken away from one group of actors and are transferred to other actors (at home or abroad, present or future). However, the distributional impacts of a particular type of instrument may be judged unacceptable from the point of view of income policy, especially if some form of corrective action (e.g., side-payments) is not feasible or practical. It is important to note that different types of instruments have different distributive consequences.

Administrative feasibility and costs

A major precondition to the effectiveness of policy instruments is that they are enforceable at reasonable costs. Efficiency does not only demand that the adjustment and implementation costs of the polluting agents are minimised but it also requires that the total costs of adjustment, implementation, enforcement and administration are minimised. Enforcement costs may therefore be an important element in the choice of instruments.

Acceptability

Opposition of target groups to a particular instrument, because it is judged 'unfair' or too complex, may raise enforcement costs considerably and may therefore negatively affect the overall effectiveness and efficiency of the instrument. According to OECD, the acceptability of instruments can be increased by: adequate and timely information on specific features of the instrument to target groups; consultation with the target groups in the execution of the instrument; and by progressive implementation, that is, polluters should be able to adjust to the new situation.

The international acceptability of the instrument is increased if the instrument is in compliance with the PPP. International competitiveness and trade may be distorted if different countries employ different environmental policy instruments for specific purposes. The extreme case is that one country would subsidise an industry to abate pollution while another country would charge this industry for the same purpose. The PPP has been adopted by OECD countries to avoid this kind of instrument-induced distortions. It demands that the polluter should bear the costs of carrying out the measures decided by the public authorities to ensure that the environment is in an acceptable state. In other words, the costs of these measures should be reflected in the costs of goods and services which cause pollution in production and/or consumption. Such measures should not be accompanied by subsidies that would create significant distortions in international trade and investment (OECD 1986). Hence, PPP is basically a 'non-subsidy' principle. Notwithstanding the non-subsidy character of the Principle, financial assistance for environmental control measures has not been ruled out in actual implementation of the Principle (e.g., by

OECD in 1972 and the European Community in 1986). Subsidies for R&D in clean technology, or in cases where the costs of pollution reduction for the industry would be disproportionally high, are regarded as not conflicting with the PPP. In OECD countries, financial assistance is widespread: they have either regarded some forms of financial assistance as PPP-compatible or as acceptable exceptions.

Finally, environmental policy instruments may affect international competitiveness and trade, and policy areas, such as economic policy, income policy and sectoral policies. In evaluating the performance of policy instruments their impacts on these policy areas should also be considered. Economic policy will consider the impacts of environmental policy instruments on the level of public expenditure, government budget, international competitiveness, and in general it will consider whether the policy instruments increase or decrease market efficiency and flexibility.[3]

2.2 INSTRUMENT CHOICE IN THEORY

It is generally acknowledged that no single type of instrument is superior in every situation (see, for example, OECD 1991). Moreover, it is accepted that different types of instruments can complement each other. Nevertheless, it may be possible to identify priority instruments in specific policy contexts. Several authors have evaluated the potential performance of different (combinations of) environmental policy instruments in specific policy contexts (e.g., Downing 1984; Bohm & Russell 1985; Baumol & Oates 1990; Bovenberg et al. 1991). Summarised below are the major findings on instruments for industrial pollution control.

Communicative instruments

Information on clean alternatives, voluntary environmental management schemes within firms, moral suasion and so forth, may be environmentally effective if the adaptation required is relatively inexpensive or if it provides benefits to the firm. These benefits may be due to a reduction in costs ('pollution prevention pays') or due to a more environment-friendly image in the consumer market. However, when there are clear economic disadvantages to firms in adjusting their behaviour, the effectiveness of purely communicative instruments is doubtful. As Downing (1984, p. 294) puts it: '. . . asking people to act in the social interest when it is against their self-interest is like asking water to flow uphill'. Bohm & Russell (1985) identify three situations where communicative instruments (moral suasion) could be best used. First, if the use of other instruments is blocked for political reasons. This could be the case if there is not yet enough societal support for legal action. Information and moral suasion can then be a first step towards more stringent policies. Second, it may be that the monitoring required for economic incentive schemes or regulation is technically infeasible or prohibitively expensive. Third, in certain cases where there is need for immediate action (e.g., extremely hazardous smog levels) ordinary instruments may be too cumbersome or too slow.

Dutch economic and environmental policies are, to a greater extent than those of the United States, based on a model of consensus between government, trade unions and industry. Therefore, it is not surprising that in the Netherlands communicative instruments play a greater role in environmental policy than in the United States. They are cast in the form of voluntary agreements or 'covenants' between government and branches of industry. Covenants typically include quantitative targets and a specified timetable. They are implicitly or explicitly backed up by the threat of regulatory action if these targets are not met voluntarily. According to Bovenberg et al. (1991) they offer industry a great amount of flexibility in meeting these targets and can therefore be efficient. The acceptability of voluntary agreements by the target groups is of course higher than the acceptability of compulsory instruments. However, environmental groups are usually very sceptical about covenants because of their lack of legal sanctions (Opschoor et al. 1994); they consider these voluntary agreements to be second-best instruments only, though preferred to no intervention at all. Important conditions for the success of covenants consist of the level of organisation and discipline within the industry-branch—to avoid free riding of individual firms; a relative stable composition of the branch—few entries and exits of individual firms; and the credibility of the regulatory threat.

Finally, communicative instruments such as information on clean technologies and abatement technologies, internal environmental management schemes within firms, and so forth, can be very useful in combination with other (economic or regulatory) instruments.

Economic instruments

Economic instruments provide economic incentives to economic actors inducing them to behave in an environmentally more appropriate or acceptable way (Opschoor & Turner 1994). In a system of *pollution charges*,[4] the administrator levies the polluter an amount of money per unit of pollution discharged. As has already been indicated, charges have been advocated by economists because, under specific circumstances, they offer both static and dynamic efficiency. The former is offered because they allow firms flexibility in adjusting their behaviour, thus allowing them to find the most cost effective solutions. Any individual firm confronted with a pollution charge can decide how much pollution to abate. It will trade-off the costs of pollution abatement investments with the benefits of paying a lesser charge. The optimal level of pollution abatement for the firm is that level where the costs of abatement of the last unit of pollution equal the unit-level of the charge. Therefore, firms with low marginal abatement costs will abate relatively more than firms with high marginal abatement costs. This guarantees cost effectiveness in abatement, that is, for industry as a whole any level of pollution abatement is achieved by least costs.

Charges carry a dynamic incentive effect too. In contrast to fixed regulatory standards, they offer a continuous incentive for industry to develop and implement more cost effective solutions. If technical progress causes control costs to fall, an emission charge will induce polluters to lower their emission levels because the marginal control costs will fall below the charge rate. The polluter will adjust his emissions until the marginal control costs are again equal to the charge rate. The impact of charges and other economic incentives on the rate and direction of innovation is largely unknown, though (Opschoor & Turner 1994).

When the administrator does not have perfect knowledge of the control cost functions of the polluting firms, the environmental effect of any charge is not known in advance. It has been suggested that the administrator adjusts the charge level via a trial-and-error approach until the desired environmental outcome is achieved

(Baumol & Oates 1990). However, the 'errors' in the charge setting may produce high compliance costs because of the huge and at least partially irreversible investments they induce (Bohm & Russell 1985, p. 410). A complicating factor is that the 'optimal' charge level can change in time because of inflation, technical progress and growth (decline) of the polluting sector(s).

The situation worsens if the *location* of the source of pollution matters and if the relationship between emissions and environmental damage is very *non-linear*. In these cases charges should be individually tailored to the various sources, and optimality seems very difficult to achieve without full knowledge of the sources' individual cost functions (Bohm & Russell 1985).

Industry's opposition to charges is to a large extent motivated by their distributional consequences. An effluent charge is levied on every unit of pollution. Therefore, even if a firm cuts down its emission to an 'optimal' level, it still pays charges for the non-abated or residual emissions. Why should this money be paid to the authorities after all? Is the state the owner of the environment? Moreover, the question is what happens with these 'transfer' payments. Will these transfers be returned to industry (for example in the form of a reduction in wage costs), or will these transfers be used for other purposes (e.g., flow to the general budget)? Industry's opposition to charges and preference for regulatory approaches can also be explained by greater rent-seeking opportunities of the latter approach (Verbruggen 1991). These issues will be pursued in Section 2.3.

The administrative feasibility of effluent charges in terms of monitoring does not fundamentally differ from direct regulation. However, the administrative feasibility could be significantly enhanced if the discharge or emission in question could be directly related to the use of some input or the production of some output. For example, in the case of a CO_2 reduction policy, a charge can be levied on carbon inputs rather than on its emissions. In this case, the charge collection is rather easy because a connection can be established with existing tax mechanisms. A clear disadvantage of this type of charging is that it does not take into account the possibility of reducing emissions by means other than reducing the amount of the charged input or output.

An alternative to charges (which set prices) is the concept of *tradable permits* (which set quantities). A permit allows a firm a certain amount of pollution per time period. Making permits tradable means that if a firm pollutes less than allowed, it can sell its surplus permit to another firm which would like to pollute more than allowed by its current permit. Any individual firm will trade-off the costs of pollution abatement investments with the benefits of selling its surplus permits. Contrary to charges, the overall level of pollution reduction because of this instrument is known in advance. There is no prior knowledge of the price of the pollution permit, it is automatically established in the market place (if trade actually takes place). The concept of tradable permits has been hailed by economists as an effective and efficient instrument for environmental policy.

However, some caveats should be mentioned. A big threat to any marketable permit scheme is market thinness. Any potential advantages of marketable permit schemes soon evaporate if there is too little trade. A possible reason for little trade is strategic behaviour of firms in an oligopolistic market. Firms may decide not to sell their redundant permits (to their competitors in the sales market) because of strategic reasons: limiting the market share of their competitors. Empirical research has shown that the major cause of market thinness is over-regulation (Dudek & Palmisano 1988). Authorities are easily persuaded to place all sorts of (well-meant) restrictions on trades. Instead of 'better' trades, the end result is often that no trade takes place at all.

Just as with charges, a marketable permit scheme gets complicated if the location of a polluting source matters. To overcome this disadvantage it has been suggested to restrict trades to sources within specific geographical zones, e.g., a particular watershed. Trades between sources in different zones would not be allowed. This solution increases the danger of market thinness and thus potentially reduces the effectiveness of the scheme.

The distributional consequences of a permit trading scheme depend on the initial allocation of the permits. Permits can be granted for free on the basis of existing pollution ('grandfathering') or they can be auctioned, whereby the administrator sells the permits to the highest bidders. Either way, the distributional consequences are controversial.

The dynamic efficiency of a permit trading scheme, i.e., the incentive to develop and implement cost effective solutions to reduce pollution, depends on the functioning of the market for permits. The incentive is greater if there is less uncertainty about the price at which the redundant permits can be sold. A high level of uncertainty on these prices (because of poor functioning markets or an unpredictable government policy) reduces the dynamic incentive.

The administrative requirements of a tradable permit scheme are higher than a comparable charge or regulatory scheme because, in addition to monitoring emissions, the control agency has to record completed trades as well.

Economists have often questioned the effectiveness of *subsidies* on clean alternatives because although they may be effective for a single firm, they do not give the right price signals to industry as a whole. They may therefore have overall perverse environmental effects because they decrease average costs and may thus lead to a higher production volume (Baumol & Oates 1990). An additional disadvantage of subsidies is that they are a burden to the general budget. Although subsidies are sometimes permitted under the various interpretations of the PPP (OECD, EC), they are often conflicting with the letter and the spirit of this Principle. A distinction can be made between direct object subsidies on the one hand, and subsidies for public provisions (e.g., collective waste water treatment plants) and subsidies for research and development (R&D) on the other hand. In general, subsidies for public provisions and R&D are less controversial than object subsidies. Moreover, (temporary) subsidies can play an important role as the 'lubricants' of environmental policy, helping firms adjust to more stringent regulations.

Deposit–refund systems imply that a potential polluter is taxed (deposit) in advance and that this tax is rebated (refunded) if certain conditions are met, e.g., proof that a product is returned to a specified place or that a specified type of damage has not occurred. Deposit–refund systems are often used to prevent littering (bottles, cans) and to prevent the uncontrolled disposal of hazardous materials such as batteries. Deposit–refund systems are attractive if it is very difficult to monitor discharges of pollution. From the government's point of view, it is an attractive option because it reverses the burden of proof. Of course, it requires that it is technically feasible and not too expensive to establish this proof. The government can implement its own deposit–refund systems or it can 'persuade' industry branches to set up their own systems.

A final economic instrument is *liability legislation*, requiring the polluter to compensate the environmental damage he caused, and thus providing a financial incentive for pollution prevention (either directly, or indirectly, through higher insurance premiums). The effectiveness of liability legislation depends on the question whether it is possible to establish proof of a cause-effect relationship between environmental damage and the discharge of a pollutant from a specific source. Examples are soil pollution by hazardous wastes and oil spills. Strict liability may force potential polluters to take the (potential) environmental damage costs of their actions into account, thereby creating incentives for efficient levels of precaution.

A general precondition for the effectiveness of economic instruments is the well-functioning of markets. The presence of severe distortions in the market in which the instruments are to be applied may diminish or nullify their effectiveness. Examples of such market imperfections are a limited access of the target group to the capital market which prevents otherwise profitable abatement investments; and the distortion of price signals which may drive a wedge between the agent who pays the charge and the agent who influences the amount of pollution (a problem known as the principal agent problem).[5]

Regulatory instruments

Regulatory (CAC) instruments have been extensively criticised by economists but are nevertheless the most common environmental instruments across the world. Section 2.3 investigates possible explanations for this phenomenon.

The disadvantages of regulatory instruments are several. First, they usually take little account of differences of pollution control costs between sources. Second, they usually apply different standards to old and new industries. Third, fixed standards offer no incentive for dynamic efficiency. Producing less pollution than allowed in the licence offers no benefits to the polluter. Fourth, a regulatory approach directed towards individual sources offers little flexibility in the face of exogenous change. Any increase in the number of regulated sources poses a threat to ambient environmental quality. A fifth disadvantage, which they share with effluent-based charge and tradable permit schemes, is that they require extensive monitoring and control.

There is little disagreement, however, that in certain specific cases of pollution there are no alternatives to direct regulation (WRR 1992), namely:

- for activities with serious impacts on health (e.g., the use of specific types of asbestos), a direct ban would be most appropriate;
- for situations which have to be solved in a very short period of time (days, weeks or months), direct regulation would be most appropriate, or, in the absence of a legal basis, voluntary agreements;
- in situations where the non-attainment of environmental targets would result in large risks or very uncertain situations (e.g., the storage of nuclear waste), direct regulation is preferred.

Moreover, the disadvantages of a regulatory approach vis-à-vis a market based approach are smaller if the location of the source matters. The smaller the disadvantages, the more stringent are the ambient environmental targets. If the targets are so strict that every source should take every conceivable measure to reach the ambient target, there will be little difference in the total cost of regulatory and market based approaches.

Standards can be divided into performance standards (e.g., effluent standards) and design standards (e.g., Best Available Technology). From the point of view of cost effectiveness, performance standards are to be preferred over design standards because they offer the firm greater flexibility in meeting the standard. Design standards can only be efficient if direct or indirect monitoring of releases of pollutants is unreliable, prohibitively expensive or technically infeasible or if there is no doubt about the most efficient approach to meet a performance standard. An example of such a design standard is the three-way catalytic converter in cars which reduces NO_x, CO and VOC emissions, because it would be infeasible to monitor exhaust emissions directly or indirectly.[6] On the other hand, this example shows the relative inflexibility of a regulatory standard in the face of exogenous change, as the obligation to use catalytic converters cannot avoid an increase in total emissions due to an increase in traffic intensity.

Fixed standards—performance or design—offer no incentive for the continuous development and implementation of cost effective solutions for pollution control. However, standards do not have to be fixed. Standards can achieve some amount of dynamic efficiency if they are progressively tightened according to some specified timetable.

Summary

The environmental policy maker has a number of instruments at his disposal. We have distinguished:

1. Communicative instruments:
   information and education;
   covenants: voluntary agreements.
2. Economic instruments:
   R&D subsidies;
   subsidies on 'clean' alternatives;
   pollution charges: a levy per unit of emission;
   product/input charges: a levy on a polluting product or input;
   tradable permits: a polluter who pollutes less (more) than his current permit can sell (buy) surplus permits to (from) other firms;
   deposit–refund;
   liability legislation.
3. Direct regulation (CAC):
   emission standards: specifying the allowable emission per time unit;
   design standards: specifying process characteristics.

Not every instrument is equally suitable in every situation. Next we will pay closer attention to the circumstances which determine the relative success of the different types of instruments.

Counter-indications of suitability

From the above discussion it is clear that specific circumstances decide on the suitability of a particular type of instrument. These circumstances are sometimes called situation–characteristics. Bovenberg et al. (1991) distinguish between three important classes of situation–characteristics:

- character of pollution;
- character of the polluting processes;
- character of the target group and the market.

Within these classes of situation–characteristics there are several aspects which affect the potential performance of policy instruments. Bovenberg et al. indicate which instruments are not, or less, suitable in the presence of specific situation–characteristics. They give counter-indications rather than prescriptions. For example, one aspect of the character of the target group is its accessibility (which increases if the target group is better organised). If the accessibility of the target group is small, the instrument of agreements (covenants) is less suitable. The situation–characteristic of 'accessibility' provides a counter-indication of suitability of covenants in this case.

Aspects of the pollution itself which are of importance for the choice of instruments include: the severity (and irreversibility) of pollution; the relationship between discharges and damage (uncertainty of cause and effect; non-linearity; location matters).

Aspects of the polluting processes which are relevant are: the availability of 'clean' substitutes or the options to develop such substitutes; the presence of market imperfections which diminish the sensitivity to price signals; the variation of abatement costs between polluting sources; the complexity of the process; and the feasibility (and costs) of monitoring polluting discharges.

Aspects of the target group and the market which are relevant are: the size and level of organisation of the target group; the market form (pure competition, monopoly, oligopoly); and the dynamics of the market (entry and exit).

The main elements of the assessment of the relationships (Bovenberg et al. 1991) between these situation–characteristics and the non-suitability of various instruments are presented in Table 2.1. The table should be read as follows. The rows represent various instrument types (covenants, R&D subsidies, subsidies on alternative products or processes, emission charges, etc.). The columns represent the various relevant situation–characteristics (severity of environmental damage, non-linearity between discharge and effect, etc.). A less suitable combination between an instrument and a situation–characteristic is marked with an X (the

Table 2.1
Counter-indications of Instruments[1]

Instrument	*Severe damage*	*Non-linearity*	*Time & space dependence*	*Few substitutes/ high costs*	*Market imperfections*	*Innovations likely*	*Innovations unlikely*	*Large variation in control costs*
Covenant	X			X				
R&D subsidies	X						X	
Subsidies on alternative	X	X	X	X[2]		X		
Emission charges	X	X	X	X[2]	X			
Product/input charges	X	X	X	X[2]	X			
Tradable permits[3]			X		X			
Tradable permits[4]			X	X[2]	X			
Liability				X[2]	X			
Direct regulation[5]								X
Direct regulation[6]	X	X				X		X
Deposit–refund		X	X		X			

Instrument	*Complex process*	*Poor measurability of emission*	*Little relation between product/input emission*	*Poor accessibility of target group*	*Little competition*	*Dynamic sector (easy entrance)*
Covenant				X		X
R&D subsidies	X			X		X
Subsidies on alternative	X					X
Emission charges		X			X[7]	
Product/input charges			X		X[7]	
Tradable permits[3]		X			X	
Tradable permits[4]		X			X	
Liability			X[8]			
Direct regulation[5]		X		X		X
Direct regulation[6]	X					X
Deposit–refund			X[9]		X[7]	

[1] counter-indications are marked with 'X'
[2] relative high distributive effects
[3] free distribution of initial permits
[4] auctioned sale of initial permits
[5] emission standards
[6] design standards
[7] charge level should take account of market form
[8] if liability is defined for a specific product
[9] deposit–refund is usually for a (rest) product

Source: Bovenberg et al. 1991.

counter-indication). Obviously, these counter-indications should not be interpreted too strictly. An instrument can be made more suitable by a careful design which takes into account the situation–characteristics.

Severity of damage

If the environmental damage is very severe and irreversible in the short term, those instruments which offer little short-term certainty on the environmental effect, i.e., covenants, subsidies, charges, design standards, seem less suitable. In Table 2.1 these instruments have been marked with an X to indicate that they may be less appropriate in cases of severe damage. This does not mean, however, that incentive instruments (subsidies, charges) cannot be used *in combination* with direct regulation. They can stimulate dynamic adjustments.

Non-linearity

In some cases environmental damage can progressively rise if total pollution exceeds some threshold level. Instruments which directly regulate the quantity of pollution are preferred in these cases. Subsidies, charges, design standards and deposit–refund instruments are counter-indicated.[7] Tradable permits and emission standards are more appropriate, although it should be noted that next to emission standards per source, the number of sources should be regulated. There is, in general, no objection to combinations of direct regulation and economic incentives.

Time and space dependency

If location and time matter, economic instruments face important difficulties, as discussed earlier. They have all been counter-indicated. If environmental damage can be relatively easily traced back to a particular source, liability can be efficient. Direct regulation can take time and space considerations into account, although this puts a heavy information burden on the environmental authority.

Few substitutes/high costs

If firms have few substitutes for dirty processes, or if they entail high costs, every single instrument will meet fierce opposition. However, because of the distributional consequences regulatory instruments (and subsidies) will face less opposition than charges and auctioned permits. Subsidies on alternatives have been counter-indicated because they will be extremely costly to the budget.

Market imperfections

As discussed, market imperfections reduce the suitability of economic instruments. Market imperfections, in general, reduce the demand–elasticity of dirty products and activities: they reduce the target group's sensitivity to price-signals. One important market imperfection is lack of knowledge. If a polluter is unaware of the possibilities of environment-friendly alternatives, a charge will not change his behaviour. Another market imperfection may be limited access to capital. Even if an environmental investment would be paying, limited access to capital may prevent the investment from being made. Market imperfections can also be created by government policies, e.g., the assistance of certain economic sectors for a variety of reasons. The first-best remedy in the case of market imperfections is to abolish them. If this is not possible, or not desirable, direct regulation is a more appropriate approach in the case of severe market imperfections. Even so, there is a real danger that direct regulation will hamper the operation of the market even more.

Innovations likely/unlikely

If it may be expected that environment-saving innovations will be developed in time, subsidies on alternatives and design standards may be very counter-productive. If innovations are unlikely, R&D subsidies are less appropriate.

Large variation in control costs

Such variation across sources usually makes direct regulations very inefficient. Large variation in control costs is *the* condition for the efficiency of economic instruments. Therefore, direct regulation has been counter-indicated.

Complex process

The information requirements of authorities to target subsidies to complex industrial processes are usually too high. In this case, design standards can be disastrous. Environmental authorities should not directly interfere with complex industrial processes because they will, as a rule, lack the required knowledge. Instruments which require less detailed knowledge of the process (permits, charges, liability) are more appropriate.

Poor measurability of emissions

If it is not feasible or prohibitively expensive to monitor emissions, any instrument which requires monitoring is less suitable: i.e., emission charges, tradable permit schemes, emission standards. Product/input charges, design standards, liability, and deposit–refund schemes offer better perspectives. The compliance of an industry with a voluntary agreement is difficult to monitor with poor measurability of emissions. Firms within an industry may also have difficulties in checking individual compliance among themselves. However, it is possible that approximate indicators can be used to check compliance in a covenant.

Little relation between product/input and emission

If there is no compulsory relationship between product/input and emissions, any instrument addressing product/inputs will be less effective. Charges on products/inputs and deposit–refund schemes have been counter-indicated. The emissions themselves should be addressed.

Poor accessibility of target group

Poor accessibility of the target group reduces the suitability of agreements and R&D subsidies. If the target group is very large and not very accessible, it will also be difficult to use emission standards since the monitoring costs may well become very high.

In our view, the same difficulty applies to the use of emission charges and tradable permits (in Bovenberg et al. charges and tradable permits were not counter-indicated).

Little competition

The market form is important for the design of economic instruments. In monopolistic markets, charges may actually increase market distortions. Marketable permit schemes may suffer from strategic behaviour and/or market thinness.

Dynamic sector (easy entrance)

In a highly dynamic sector, both agreements with branch organisations and source-specific direct regulation are less suitable. The first is less suitable because of the free-rider phenomenon, the second because of the fact that source-specific direct regulation is not flexible in the face of exogenous change. Emission standards per source will not control the total pollution load to the environment if the number of sources grows significantly. Tradable permits which put a cap on total pollution seem to be the preferable instrument in this situation.

Private versus public enforcement

In most countries the responsibility for environmental regulation and control has been vested almost exclusively in public agencies. With the number and complexity of environmental risks growing, it is increasingly realised that these agencies face limits with regard to their capabilities. For this reason, recent research has focused on the complementarity and substitutability of private and public action in the field of environmental regulation and control (Tietenberg 1992b; WRR 1992). The role of the court system in pollution control has received considerable attention, especially since liability law has been forwarded as a major instrument in two US environmental Acts.[8] Liability law can be an efficient instrument in some cases if it is not too difficult to establish proof of a cause-effect relationship between environmental damage and the discharge of a pollutant from a specific source. Examples are soil pollution by hazardous wastes and oil spills. Strict liability may force potential polluters to take the (potential) environmental damage costs of their actions into account, thereby creating incentives for efficient levels of precaution (Tietenberg 1992b). However, much depends on the actual characteristics of the liability law.

Another innovation which may reduce the burden of public agencies is the obligation of corporations to regularly prepare and publish environmental accounts, in the same vein as they are forced to prepare and publish annual financial accounts. A 'green' image is becoming increasingly important for companies servicing the consumer market. Voluntary 'green' or 'eco' labels could be issued on the condition of published and approved environmental accounts.

The Dutch Scientific Council of Government Policy (WRR 1992) has systematically distinguished the options of government to influence the behaviour of economic agents with respect to the environment. We have already discussed the distinction between communicative instruments, economic instruments, and direct regulation (CAC). In more general terms, WRR calls these instruments persuasion, transaction and coercion. A second dimension is formed by the capacity in which the government acts: in the design and execution of public law, in the design of rules of private law, and as a 'normal' participant in societal interactions. The final role of government is as a provider of infrastuctural and complementary facilities which are meant to broaden the range of behavioural alternatives of economic agents.

The different options and roles of government are summarised in Table 2.2.

2.3 THE CHOICE OF ENVIRONMENTAL POLICY INSTRUMENTS IN PRACTICE

In the preceding section, we have identified a number of situation–characteristics which contribute to, or reduce the applicability and feasibility of specific policy instruments. It can be concluded that in theory there are many instances where the use of economic instruments should be preferred to direct regulation. The question then naturally arises: why is it that environmental policy makers throughout the world, in the vast majority of cases, opt for a CAC approach towards pollution? To find some possible explanations for this phenomenon, it may be useful to have a closer look at the

Table 2.2
Government's Options to Influence the Behaviour of Economic Agents with respect to the Environment

coercion	public law	private law	participant
	direct regulation		
transaction	charges, subsidies to polluters	contracts, liability law	
persuasion	environmental accounting, environmental labelling, subsidies to environment organisations		gentleman's agreements, information, education

Source: WRR 1992.

specific interests of the various stakeholders involved in the process of environmental policy making: the authorities, the polluting firms, environmental pressure groups, trade unions and consumer organisations.

The *policy makers and administrators* are, supposedly, interested in achieving quick and predictable results in terms of environmental improvement. It is often assumed that a CAC type of policy matches this goal best, as it implies enforceable targets. However, in practice it often appears to be hard to enforce standards, bans and obligations effectively. One could even argue that economic instruments (especially charges and taxes) have more of a built-in incentive for effective enforcement because a government which does not enforce them loses money. Also, in fiscal affairs the principle of 'equal treatment of equal cases' seems to be more established than in environmental affairs. In other words: it seems to be less acceptable to prosecute one tax evader and spare the other, than to treat two offenders of environmental standards differently. Nevertheless, environmental policy makers often seem to be convinced of the superiority of direct regulation in terms of effectiveness.

Another reason why policy makers prefer direct regulation might be that they are afraid of having less grip on polluters if they would apply charges instead of standards (Opschoor & Turner 1994). Many authors also mention 'bureaucratic inertia' as a cause: the fact that administrators are more familiar with regulatory than with fiscal measures and procedures. Hanley et al. (1990) add to this the fact that vested interests are involved: the CAC approach creates more jobs and influence for administrators than allowing the market to perform some of the regulation. They also mention the relatively weak influence of economists on the formation of environmental policy. In line with this, environmental administrators are afraid to lose their autonomy if economic instruments are being applied. Taxes and charges generally imply the involvement of the finance department, whereas direct regulation is the primary or even sole responsibility of the environmental department.[9] On the other hand, the finance administrators are often quite reluctant to introduce environmental taxes and charges because they are mainly interested in a stable revenue and a simple tax base, two features which effective pollution taxes will not display principally.

The individual *polluters'* most preferred instrument is, obviously, the subsidy. In most industrialised countries, however, industry has reconciled itself to the PPP which implies that subsidies for pollution reduction are not allowed in principle. Probably, the 'second best' instrument for industry is the voluntary agreement, provided it can be ascertained that no 'outsiders' take advantage by circumventing the provisions of the agreement.

When it comes to the choice between direct regulation and economic instruments, both industrial organisations and individual enterprises are generally in favour of direct regulation. Several reasons can be identified for this.

Buchanan & Tullock (1975) have shown that under certain conditions regulation is less costly to the firm than the pollution tax alternative. Although their analysis applies only to a special case, it is indeed true that the burden of environmental taxes, despite the efficiency advantage for society at large, may be heavier for the polluter than the direct regulation with environmentally equivalent effects.[10] In principle, the revenues of a pollution tax can be returned to the polluters (on a basis which should be independent from the amount of pollution), but in actual practice the money mostly flows to the (general or environmental) government budget. Therefore, it is understandable that industry is not in favour of the tax alternative.

A second set of reasons for industry to be in favour of regulation can be lumped together as 'rent seeking' (Verbruggen 1991). Direct regulation can provide industry with several kinds of advantages and market protection. This is evident in cases where 'newcomers' have to meet more stringent requirements than established firms.[11] Furthermore, standards are often formulated in such a way that they can only be met by a specific technology, thus favouring the supplier of that technology. Indeed, one can observe many instances of producers trying to obtain the status of legal standard for their 'clean' product or process. Generally speaking, one could say that regulation in general (and also environmental regulation) may serve protectionist interests in various ways. Economic instruments are more transparent and less vulnerable for serving protectionist interests, and may thus have a more direct influence on the industry's international competitiveness (Verbruggen 1993).

Third, standards are more susceptible to bargaining and negotiations than charges and taxes. In many countries, the essentials of a tax (including the rate) have to be laid down in the law whereas standards often leave much to the discretion of the environmental administrator. Industry can try to use their influence and get more lenient standards by showing that the proposed standards are technically or economically not feasible. This can be done both at the level of the individual firm and by industrial organisations.[12] On the other hand, in case of a pollution tax the industrial lobbies can only try to influence the political process leading to fiscal legislation. Once the tax law is in force, they can only choose between paying or evading the tax (or bribing the tax officials, if their ethics permit them to do so).

Environmental pressure groups, presumably acting on behalf of the (potential) victims of pollution, are traditionally opposed to replacing direct regulation by fiscal instruments. They dislike the idea of 'selling' the environment to the highest bidder (thereby forgetting that in a system of standards, the polluters are granted free use of the environment). They somehow believe that the right to pollute should not be for sale because the value of the environment cannot be expressed in money terms. In recent years, however, the attitude of environmentalists has shifted more in favour of pollution charges and taxes, although economic instruments are still not seen as an alternative, but more as a complement to direct regulation. Many 'green' political parties and organisations have advocated a radical tax reform, shifting the tax burden from labour to polluting products (rather than emissions).

Finally, the citizens also play other roles on the environmental policy stage. The organisations which represent them in these roles (for instance, *trade unions* and *consumer organisations*) are generally not in favour of any charges or taxes which decrease their purchasing power or lead to higher prices of consumer products. On the other hand, in many countries these organisations tend to be increasingly prepared to make sacrifices and moderate their demands in exchange for concrete environmental policy measures. When it comes to the choice between instruments, consumers will probably be more opposed against product charges (with clear and direct effects on product prices) than against measures which initially hit the producers only (such as direct regulation or emission charges).

Where economic instruments *have* been introduced in practice, the main reason for doing so was generally the need for revenues and not their efficiency advantage with respect to direct regulation. Most of the existing charges and taxes were introduced in order to finance certain collective environmental facilities (Opschoor & Vos 1989). In some cases, these financially motivated charges have had a clear regulating side effect, by stimulating polluting firms to save costs by carrying out pollution control activities themselves. The Dutch water pollution fee is a well-known example (Bressers 1988).

Another practical reason for introducing economic instruments is their flexibility. In the United States, marketable permits have offered a solution for the problem that in the case of fixed standards no expansion of production is possible in the so-called 'non-attainment areas'. Under a marketable permit system, the producer can obtain the right to increase emissions (caused by an increase in production) by buying excess emission rights from other firms.

Most of the existing economic instruments are being used in addition to CAC types of instruments (Opschoor & Vos 1989). In some cases, however, they came into being as a substitute for direct regulation, especially when a standard or an outright ban was deemed unfeasible by the environmental authorities. Many EU countries have, for instance, stimulated the use of 'clean' cars and lead-free petrol by means of tax advantages because EU regulations did not allow a ban on 'dirty' alternatives.

Thus, it can be concluded that several reasons can be found for the fact that the role of economic instruments in environmental policy is still relatively modest. We assume that many of these reasons lie within the area of the interests of major stakeholders involved: policy makers, administrators and firms. Nevertheless, many experts foresee a larger role for market based instruments in future environmental policy. Hahn (1989) mentions three factors which stimulate their application:

- the increase in marginal abatement costs as environmental standards are tightened;
- the experience gained in implementing economic instruments;
- the decrease in monitoring costs.

One could add to this the fact that economic instruments have a certain appeal to free-market supporters and seem to reduce the need for regulation. Also, the need for tax reforms (or additional tax income) may lead in some countries to increased attention for environmental charges and taxes, as has been the case in Sweden (a shift from direct to indirect taxes). Finally, for some pressing environmental problems there seems to be no feasible alternative to economic instruments. This is for instance the case with global warming (partly caused by CO_2 emissions). Curtailing energy use by means of direct regulation (rationing?) seems to be too rigid, while trying to do so by communicative means has proven to yield insufficient results. Therefore, economic instruments (charges or tradable rights) remain as an option.

2.4 CONCLUSIONS AND ANALYTICAL FRAMEWORK FOR THE REST OF THIS STUDY

Direct regulation (CAC) has been and is the most important instrument in environmental policy across the world. Economists have often advocated a wider use of economic incentives to make pollution control more cost effective. It has also been realised that the effectiveness of direct regulation may be threatened by enforcement problems in the face of the ever-increasing number and complexity of environmental risks. This has led to a search for possibilities to reduce the burden of monitoring and control by public agencies, for example, through the obligation of environmental accounting or through liability law.

There can be no general statement about the desirability of any single type of instrument in environmental policy. Section 2.2 has considered many practical complications influencing the choice of instruments in specific situations in a particular country. The situation can differ with respect to three different aspects. The first is the environmental problem: the severity of the damage, the relationship between discharges and damage, and whether location matters. The second is the character of the polluting products and processes and their alternatives: the availability and costs of clean substitutes, the ease of monitoring discharges, the variation of abatement costs between sources. The third aspect is the character of the market: market form, market imperfections, the size and level of organisation of the target group. Section 2.2 provided us with a preliminary framework to assess the suitability of various instruments in specific circumstances.

In this study, the performance of pollution control instruments in different countries is compared. Each country has its own value judgements regarding the acceptability of certain instruments and their (distributional) consequences. An important difference may lie in the organisation, competence and capabilities of public pollution control agencies, on national, state and local levels. One should also consider the present judicial system for assessing its potential role in complementing public environmental policies. Finally, one should have an open eye for possible innovations in enforcement and the role of NGOs and local communities therein. We, therefore, identify the institutional, country-specific context as a fourth aspect which can influence the choice of instruments.

Keeping these differences and practical complications in mind, there seems little objection in stating that in the search for an effective and efficient set of environmental policy instruments one should always try to design a system 'wherein the self-interest of the individuals will cause the emitters to control efficiently, the recipients to seek and enjoy efficient environmental quality, and the agency personnel to be rewarded for helping produce it' (Downing 1984).

This book endeavours to evaluate the suitability of the above framework for a typically non-western economy and tries to expand the framework with typically country-(and development) specific situation–characteristics. In doing so, this study tries to answer the following questions:

- Can a more extensive use of economic instruments contribute to a more effective and efficient environmental policy in both countries, given their respective situation–characteristics?
- To what extent are differences in pollution control objectives and instruments in India and the Netherlands justified by differences in economic and other conditions?

To deal with these questions, it is necessary to have adequate information on the environmental policy and its context in India and the Netherlands. The purpose of Part II of this book is to provide this information.

Notes

1. As has been mentioned already in Section 1.2, the idea of a tax to internalise externalities originated from A.C. Pigou in his *Economics of Welfare* (1920). Taxes or charges for this purpose are often called *Pigovian* taxes, after their inventor. The fact that taxes offer a lower-cost method of achieving a given standard than CAC was formalised by Baumol & Oates (1971). The idea of tradable pollution permits was introduced by J.H. Dales (1968).
2. Subsidies usually introduce such distortions. The negative environmental consequences will be discussed later.
3. For an extensive discussion on the interplay between environmental policy instruments and economic policy, see Bovenberg et al. 1991.
4. The following discussion refers to a system of charges which has the primary objective of reducing emissions (so-called incentive charges). As we will see in Section 2.3, in practice the main objective of pollution charges is usually to raise revenues. Rates are often too low to provide any incentive effect.
5. This problem is manifest in the case of protected rents in the residential sector. Although it might be profitable for both landlord and tenant to better insulate the house to save energy, the incentive on the part of the landlord is lacking if he is unable to pass on the insulation costs to the tenant because of the protected rent.
6. It cannot be based on fuel use because the emissions are dependent on the actual combustion process.
7. The disadvantages of charges can be mitigated if they are combined with other measures which ensure compliance with the desired standard, e.g., a prohibitively high charge rate for pollution levels exceeding this standard, or a rebate scheme for firms which achieve abatement beyond this level.
8. The Comprehensive Environmental Response, Compensation and Liability Act of 1980, and the Oil Pollution Liability Act of 1990.

9. From this point of view, environmental administrators could be expected to prefer, within the area of economic instruments, marketable permits to taxes or charges.
10. Even if the tax option leads to lower costs for industry than direct regulation, it is sometimes argued that industry prefers regulation because its costs are less 'visible'. In any case, the costs of regulation do not increase the tax burden figures whereas environmental taxes and charges generally do so.
11. Cairncross (1991, p. 239) cites several examples of such 'dirty protectionism'.
12. Clearly, if an individual firm succeeds in obtaining less stringent environmental demands than its competitor, this again creates an artificial 'rent'.

Part II

COUNTRY STUDIES

3

INTRODUCTION

To achieve environmental harmony, environmental problems are being assigned top priority by almost all the countries irrespective of their level of development. However, policy instruments used to achieve the objectives vary with the level of development. The differentials in the policy approach have to suit the variations in magnitude and sources of environmental problems. For efficient environmental management, it has been argued that there should be flexibility in the application of environmental regulations even between the developed and less developed regions of a country (Shane 1979).

While a detailed assessment of the environmental policy adopted by a single country would reveal the problems, prospects and more specifically the changes required for efficient environmental management, a comparative study between two countries would enable both the countries to set right their policy approach by learning from each other's experiences.

The main problem that one finds in this study is the comparison between the unequals. The Netherlands is a highly developed economy, while India is a developing economy with a long way to go in economic development. This has been supported by some basic indicators of the economies of the two countries (Table 3.1). However, both are democracies, though India has a federal constitution with a strong centre and the Netherlands has a unitary set-up. They are also very unequal in size. The study has tried to address this issue by its focus on a single state in India: Karnataka. But the main point on which this study is made valid is the similar policy instruments adopted by both the countries to combat industrial pollution—direct regulations supported by economic

Table 3.1
Some Basic Indicators of Economy—1981 and 1991: Karnataka, India, and the Netherlands

	Variables	Karnataka		India		The Netherlands	
		1981	*1991*	*1981*	*1991*	*1981*	*1991*
1.	Location controls						
	Latitude	Between 11° 30' and 18° 30'		Between 8° and 37°		Between 50° 45' and 53° 52'	
	Longitude	Between 74° and 78° 30'		Between 68° and 97° 30'		Between 3° 21' and 7° 13'	
2.	Area in sq.km	191,791	191,791	3,287,263	3,287,263	41,548	41,547
3.	Population	37,135,714	44,806,468	683,329,097	844,324,222	14,208,000	15,010,445
4.	Population growth (compound rate p.a. 1981–91)	1.89		2.14		0.55	
5.	Population density (persons/sq.km)	194	234	216	267	342	361
6.	Sex ratio (number of females per 1000 males)	961	960	934	929	1,016	1,023
7.	Infant mortality rate (per 1000 live births)	69*	80* (1988–89)	110	94 (1988–89)	15	16
8.	Per cent of workers to total population	40.24	41.83	36.77	37.64		42.92
9.	Net state/national product at current prices in million US $**	6,289	8,502	123,947	196,859	113,344	229,567

10.	Per capita net state/ national product (in US $)** current prices)	171	189	182	202	7,976	15,294
11.	No. of registered factories per million population	145	127	141	122	3,575	3,552
12.	No. of registered factories per hundred sq.km	3	3	3	3	122	128
13.	Per capita electricity consumption (in kWh)	148	243	135	236 (1989–90)	4,304	4,971
14.	Per cent share of power consumption in industry	74	50	58	45	45	43
15.	Per cent share of domestic consumption of power	12	16	11	17	27	23
16.	Per capita consumption of petroleum products (in kg)	34	53	47	65	1,534	1,202
17.	Per capita fertiliser consumption (in kg)	9	18 (1989–90)	8	13 (1989–90)	49 (1980)	39(sum of N, P_2O_5 and K_2O (1990)

Table 3.1 (contd)

	Variables	*Karnataka*		*India*		*The Netherlands*	
		1981	*1991*	*1981*	*1991*	*1981*	*1991*
18.	Road length per 1000 sq.km	513	671	454	620	2,612	2,845
19.	Road length per 100,000 population	266	288	218	241	764	788

* The two figures are not comparable due to methodological problems.

** Exchange rates used: 1 US $ = Rs 8.93 = 2.50 Dfl. in 1981.
1 US $ = Rs 24.52 = 1.87 Dfl. in 1991.

Note: Area, population, workforce and some demographic indicators of 1991 Census are provisional.

Sources: 1. Primary Census Abstract, General Population: India, 1981, Series I, Part II B (i), pp. 4–15, Census of India, 1981.
2. Provisional Population Totals: Rural-Urban Distribution—India, Series 1, Paper 2, Census of India, 1991, pp. 93 and 94.
3. Economic Survey: 1991–92, Ministry of Finance, Government of India, New Delhi, pp. S–3, S–34, S–100–101.
4. Year Book: Family Welfare Programme in India 1985–86 and 1989–90, Ministry of Health and Family Welfare, Government of India, New Delhi, pp. 101–2.
5. Provisional Population Totals: Workers and their Distribution, Series 1, Paper 3, Census of India 1991, pp. 61 and 63, 193 and 195.
6. Economic Survey 1990–91, Planning Department, Government of Karnataka, p. 13.
7. Annual Survey of Industries: 1987–88 and 1980–81, Summary Results for the Factory Sector, C.S.O., Ministry of Planning, Government of India, New Delhi. 1980–81: pp. 63 and 79; 1987–88: pp. 126 and 148.
8. Statistical Abstract: India, 1982, C.S.O., Ministry of Planning, Government of India, New Delhi, p. 131.
9. Quarterly Bulletin of Economics and Statistics: Karnataka, Directorate of Economics and Statistics, Government of Karnataka, 1980–81 and 1990–91.
10. *Historical Atlas of South Asia*, ed. J.E. Schwartzberg, University of Chicago Press, 1978, p. 78.
11. Current Energy Scene in India: Economic Intelligence Service, CMIE, May 1992, pp. 2.1, 4.21.

12. Fertiliser Statistics: India 1989–90, pp. I–78, I–86 and I–93.
13. Socio-economic Indicators: 1960–91, Directorate of Economics and Statistics, Government of Karnataka.
14. Socio-economic Trends, Karnataka: 1955–91, Directorate of Economics and Statistics, Government of Karnataka, p. 17.
15. Statistisch jaarboek (Statistical Year Book), several years, Central Bureau Voor de Statistiek (Central Bureau of Statistics), Voorburg/Heerlen.
16. *World Development Reports* 1983 and 1993.

instruments (Government of India 1992, p. 2; Amelrooy 1994, p. 28). With a more developed market economy, the Netherlands has been making effective use of economic instruments. Both countries are environmentally conscious and have strong environmental lobbies.

In the present part, the research findings have been presented under four broad headings: (*a*) Introduction, (*b*) Policy instruments for pollution control in India, (*c*) Policy instruments for pollution control in the Netherlands, and (*d*) Case studies. Under the environmental policy instruments, the study discusses separately for both the countries, historical overview and backgrounds, institutions and instruments for industrial pollution control, effectiveness of past and present policies, costs of compliance and enforcement, and prospects. To understand and compare the problems of pollution, instruments used to abate pollution, cost involved in pollution abatement and the future perspectives of pollution control for three industrial types: cement, textile and fertiliser, have been studied in both the countries and presented under the common section: case studies.

4

POLICY INSTRUMENTS FOR POLLUTION CONTROL IN INDIA

4.1 HISTORICAL OVERVIEW AND BACKGROUND[1]

With an estimated population approaching 900 million, India ranks second in number of inhabitants among the countries of the world. Its population density is also fairly high, although still lower than in the Netherlands (see Table 3.1). Agriculture still accounts for 70 per cent of Indian employment, but industry has become an important part of the economy. India has for long been following an industrial policy of state intervention and import substitution. Recently, however, liberalisation processes have started and international trade is increasing.

India is a federal republic, comprising 25 states and seven union territories. There is a strict division of tasks between the states and the federal government. Policy areas on the 'State List' (including agriculture and industry) are the exclusive responsibility of the states. Other areas are jointly administered by the two levels of government. Environment is among these 'Concurrent List' policy areas.

Environmental concern in India rests not so much on aesthetic and health grounds as on survival and livelihood issues. Though aesthetic and particularly health considerations are important, it is the survival and livelihood issues which lend even greater urgency to environmental policy than in the developed countries. On the other hand, in so far as economic growth creates livelihood opportunities, there is a tendency to weigh the benefits of economic growth in terms of such opportunities generated against the adverse impact on the environment which economic growth may create. The perception that benefits of economic growth far outweigh its adverse environmental impact has to be moderated by the fact that the people benefiting may be quite different from the people losing. The loss of the latter cannot be off-set by the gain of the former. This requires that prevention of adverse impact is built into the design of the development projects, or that the losing people are adequately compensated or rehabilitated. Because of the advanced levels of income and technological development, such a task is more easily accomplished in the North than in the South. When the North was in a comparative stage of development as today's South, it could more easily externalise the costs of development as environmental concern then was minimal. It is essential to bear this in mind before making any comparative assessment of environmental policy of a developing country with a developed country.

The primacy of development was emphasised by Mrs Indira Gandhi when in her address at the United Nations Conference on Human Environment at Stockholm in June 1972 she asked, 'Are not poverty and need the greatest polluters? . . . How can we speak to those who live in villages and in slums about keeping the oceans, the rivers and the air clean when their own lives are contaminated at the source? The environment cannot be improved in conditions of poverty. Nor can poverty be eradicated without the use of science and technology' (as reproduced in Rosencranz et al. 1991, p. 39).

Yet, Mrs Gandhi spared no pains in taking several steps, as shown later, in the case of environmental improvements. Apart from the concrete legislative and institutional measures, the need for environmental protection was explicitly incorporated into the Constitution by the Constitution (42nd Amendment) Act of 1976. The Amendment also sought to evolve a national policy and institutional framework for implementing the policy by including 'forests' and 'protection of wild animals and birds' in the concurrent list where both the central and state governments had a say; these categories were earlier in the exclusive domain of the states. Article 253 of the Constitution was added to empower the national parliament to make laws for the whole country including subjects in the state list, i.e., those under the jurisdiction of the states as per the Constitution. This provision was made use of later to make several laws in the interest of environmental regulation and improvement.

To serve as a guideline to the government, Article 48A was added to the Directive Principles of State Policy in 1976 which states: 'The State shall endeavour to protect and improve the environment and safeguard the forests and wildlife of the country'. In a new chapter entitled 'Fundamental Duties', Article 51A(g) imposed a similar responsibility on every citizen 'to protect and improve the natural environment including forests, lakes, rivers and wildlife and to have compassion for living creatures . . . ' (as quoted in Rosencranz et al. 1991, p. 53).

The Directive Principles are increasingly being cited by judges as complementary to Fundamental Rights. The courts have been guided by these Principles in several environmental cases. The Supreme Court of India has held: 'whenever a problem of ecology is brought before the Court, the Court is bound to bear in mind Article 48A of the Constitution . . .' (as quoted in Rosencranz et al. 1991, p. 54).

The commitment to environment protection and improvement was further consolidated in the 1980s and expressed on the eve of the Rio Conference in 1992 in the form of two policy statements of the government—'National Conservation Strategy and Policy Statement on Environment and Development' (June 1992) and 'Policy Statement for Abatement of Pollution' (February 1992) both released by the Ministry of Environment and Forests, Government of India. The former statement classifies environmental problems in India into two broad categories:

1. those arising as negative effects of the very process of development; and
2. those arising from conditions of poverty and underdevelopment (p. 2).

The statement expresses resolve to meet both problems squarely. To achieve this objective of sustainable development, the policy statement mentions the following instruments for action:

- to carry out environmental impact assessment of all development projects right from the planning stage and integrate it with their cost-benefit considerations. Appropriate costs of environmental safeguards and regeneration would continue to form an integral part of the projects;

- to ensure that all projects above a certain size and in certain ecologically sensitive areas should require compulsory prior environmental clearance;
- to incorporate environmental safeguards and protection measures in policies, planning, site selection, choice of technology and implementation of development projects like agriculture, water resource development, industry, mineral extraction and processing, energy, forestry, transport and human settlements;
- to encourage research, development and adoption of environmentally compatible technologies; and to promote application of the modern tools of science and technology for conservation, bridging of large gaps in supply and demand as well as control and monitoring of natural resources;
- to elicit and ensure participation of people in programmes for environmental improvement and for integrating the environmental concerns in planning and implementation of development programmes;
- to create environmental consciousness through education and mass awareness programmes;
- to aim at moderation of process of demand unleashed by the developmental process itself by taking measures to recycle waste materials and natural resources, conserve energy, conserve use of natural resources in industrial products by measures like wood substitution and generally try to reach moderations in lifestyles consistent with sustainability and human dignity;
- to develop appropriate organisational structures and a pool of professional manpower to serve as the cadre for environmental management service; and
- to effectively implement the various environmental laws and regulations for environmental protection through creation or strengthening of the requisite enforcement machinery (Government of India, June 1992, pp. 11–12).

Industrial pollution has been an important cause for environmental concern in so far as it deprives the livelihood of people through depletion of fisheries and increased incidence of morbidity and consequent loss of income and employment among the poor due to pollution of drinking water and air and higher exposure to hazards. Even agricultural yields are known to have declined due to air pollution especially by cement factories. The health of livestock is similarly affected both by air and water pollution. The Policy Statement for Abatement of Pollution lists different areas where the pollution problem is grave and shows that it needs to be tackled on a priority basis (Government of India, February 1992, pp. 2–4).

At least on paper and in law, if not always in practice, it is recognised that the victims of pollution have a right on their environment and that the polluter has to pay for environmental costs. This recognition was expressed mainly in the tort law, whereby the affected people could approach the courts and initiate legal proceedings against polluters. This remained the major approach to environmental regulation till the early 1970s. It was only later that important institutions were created which could act on their own, without waiting for the victims of pollution to initiate proceedings. Though fiscal incentives are also tried and are receiving increasing emphasis as we shall see, administrative regulation remains the main instrument of policy.

India's environmental concern with regard to pollution can be traced back to 1853 when the Shore Nuisance (Bombay and Kolaba) Act was passed. The Indian Penal Code Act of 1860 was a more comprehensive legislation containing several provisions for pollution control. Fines and even imprisonment were instituted for 'fouling of water of public springs or reservoir', for 'making atmosphere noxious to health', and for negligent conduct with respect to poisonous substances, combustible matters and explosive substances. The Police Act, 1861 dealt with noise pollution in public places. The Indian Easement Act, 1882 protected riparians against 'unreasonable' pollution of water by upstream users. The Indian Fisheries Act of 1897 penalised the killing of fish by poisoning or by the use of explosives. There were state level Acts to control air pollution also, such as the Bengal Smoke Nuisance Act of 1905 and Bombay Smoke Nuisance Act of 1912. The Motor Vehicles Act of 1939 empowered state governments to regulate emission of smoke and other nuisance (for more details, see Rosencranz et al. 1991, pp. 37–38; and Lal's 1990, pp. 402–3, 409, 448–57).

The approach to environmental regulation was piecemeal and was based on the tort law. The action against the polluters was taken only by courts on the basis of proper representation by the affected people. The penalty provided was not deterrent enough, and the course of legal action took a long time. Environment related legislation was not accorded priority, and no special institutions were created to monitor such legislation and prevent adverse impact on the environment.

This situation continued after Independence up to 1972. There was, however, legislative activity between 1947 (the year when India won Independence) and 1972 which related to pollution control among other things. The prominent steps were: The Factories Act, 1948; the Industries (Development and Regulation) Act 1951; The River Boards Act, 1956; the Atomic Energy Act, 1962; the Insecticides Act, 1968; the Merchant Shipping (Amendment) Act, 1970, and the Radiation Protection Rules, 1971. These Acts often dealt only incidentally with pollution and proved ineffective in handling it. Pollution of rivers reached alarming proportions in several places, and the Acts remained on paper. There was no institutional set-up to implement and monitor environmental regulation, and there was also a growing feeling that the whole approach was piecemeal and sporadic.

A breakthrough came in 1972 in the wake of the Stockholm Conference. The preparatory committee for this conference which identified areas of environmental concern became the precursor of the National Committee on Environmental Planning and Coordination (NCEPC) which was set up in 1972. It had the responsibility of reviewing policies and programmes for the environment. It also undertook the appraisal of several, if not all, development projects from the environmental angle and made suggestions for their modification to avoid adverse impact. Mrs Indira Gandhi, the then Prime Minister, lent her political and moral support for policies and measures to preserve the environment without of course having to sacrifice the goal of economic development and poverty alleviation. At that time, however, there was no legal requirement for environmental impact assessment of development projects. It was the political and moral influence of the government which induced it, though selectively on a few major cases. It was significant that a beginning was made in this regard. This also led to a new surge of legislative activity and institution building for environmental regulation as will be discussed shortly.

The early 1970s witnessed a milestone in environmental action in the form of the enactment of the Water (Prevention and Control of Pollution) Act of 1974. Though there were earlier several sporadic and piecemeal attempts to control water pollution as noted above, the 1974 Act was the first serious attempt at controlling it. The Act was the culmination of over a decade of deliberations between the central government and the states. The states passed resolutions permitting the parliament at the centre to pass this Act so that a uniform code and organisational structure evolved. Under this law, Boards for Prevention and Control of Pollution of Water were established both at the central and state levels with the necessary technical competence and legal power to monitor the implementation of the law. Without waiting for the affected people to launch legal action, the Boards could on their own initiate proceedings against those individuals and firms who infringe the law. A chain of water-testing laboratories was established to check water quality and pollution. The Boards could lay down standards for discharge of effluent and take action to implement them. Factories or persons are required to obtain consent to discharge effluents which can be rejected if they do not maintain the standards prescribed. The Boards can also advise on appropriate sites for locating new industry. The Boards are autonomous in nature, headed by the Chairman. There are no elected representatives on the Board, which therefore essentially remains a technocratic–bureaucratic organisation though there are also some non-official members on the Board nominated by respective governments.

To meet the expenses of the Boards, the Water Cess Act of 1977 was passed requiring industries to pay a cess on their water consumption. The Act also gave a rebate of 70 per cent of the cess in case pollution treatment plants were installed and pollution control standards were met by the industries. In 1991, this rebate was reduced to 25 per cent in the interest of resource conservation. The water cess, however, has been adequate neither in meeting the expenses of the Boards nor in acting as an effective economic incentive in inducing the firms to take up pollution abatement measures because of its nominal incidence. It is cheaper to pay the full cess than to install pollution control plants. The regulatory power of the Pollution Control Boards has therefore remained the major instrument of environmental policy.

The next important step was the setting up of the Department of Environment (DOE) in the central government in 1980. This followed the recommendations of a committee headed by N.D Tiwari (Government of India 1980). The DOE coordinates all environment programmes, reviews and clears development projects from an environmental angle, and directs policy and administrative responsibilities concerned with environmental laws. By 1985, the Department was transformed into a full-fledged Ministry of Environment and Forests. Following again the recommendations of the Tiwari Committee, NCEPC was replaced by a new body called National Committee on Environment Planning (NCEP) in April 1981. Some of the important functions of NCEPC had been taken over by the Department of Science and Technology and then by the new DOE. The main task of NCEP was to establish a nationwide Environmental Information and Communication System to propagate environmental awareness (Rosencranz et al. 1991, pp. 44–45).

It was not until 1981 that steps were taken to control air pollution. This was done through a separate Act—the Air (Prevention and Control of Pollution) Act of 1981. To enable an integrated approach to pollution control, the Water Pollution Control Boards were authorised to deal with air pollution besides water pollution. (Hence they are now just called Central/State Pollution Control Boards). The states and state level boards are required to prescribe and enforce emission standards for industry as well as automobiles after consulting the Central Board and noting ambient air quality standards. The automobile emission standards are monitored and implemented by the Commissioners for Transport and their staff in the respective states.

The approach of the Boards was judicial. The Acts were administered through criminal prosecutions initiated by the Boards and through applications to magistrates for issuing indurations to restrain pollution. This took a long time and often the polluters got away with mild penalties which did not prove to be a deterrent.

The Bhopal Gas Tragedy which took place after the midnight of 3 December 1984, proved to be a turning point in the evolution of environment policy in India. Forty tonnes of highly toxic Mèthyl Iso Cyanate (MIC) gas stored in the pesticide plant of Union Carbide located in Bhopal city leaked into the atmosphere killing around 3,500 people and seriously injuring about 200,000, a good proportion of them permanently. This led to a further spate of legislative activity and the tightening up of environmental law and implementation.

One of the first steps was to reconstitute and upgrade the Department of Environment (DOE) and transform it into a Ministry of Environment and Forests (MEF). Its status improved significantly compared to the earlier back-bench position of the DOE. The states have also constituted their own ministries or departments of environment. So far, 19 out of 25 states have done so. The states are free to apply their own (but stricter) standards than indicated by the Central Ministry or Central Pollution Control Board (CPCB). The overall coordination at the national level is done by the MEF. It takes over direct responsibility in the case of states which do not have their own ministry or department of environment.

An 'Umbrella' Act was also enacted by the national parliament entitled, The Environment (Protection) Act 1986. This Act empowers the central government to take all necessary measures to protect and improve the quality of environment and to prevent, control and abate environmental pollution. The Act identifies the MEF, Government of India, as the apex nodal agency to deal with environmental problems of the country so that an integrated and holistic policy can be implemented with regard to the environment. Unlike the earlier Acts, the scope of Environment (Protection) Act is broad, covering water, air, land and the inter-relationships that exist among water, air, land and human beings and other living creatures. That is how the MEF came to be given a key role as nodal agency. The main functions of MEF are:

1. Coordination of the activities of various central and state authorities established under the previous Acts;
2. laying down emission/effluent standards for various industrial activities;
3. powers to get information about the industrial process and to inspect the plant, premises, records and other necessary material, including seizure of documents;
4. giving directions for closure, prohibition or regulation of industrial processes as necessary;
5. stoppage or regulation of the supply of water and electricity or any other services to industries violating pollution standards.

It is not necessary for the MEF to perform these functions directly. Institutions like the Pollution Control Boards (PCBs) are empowered to act on its behalf, with overall control vested with the MEF. The necessary amendments were made to Water and Air Pollution Prevention and Control Acts to empower PCBs in this regard. The Air Pollution (Amendment) Act, 1988 sought to bring noise pollution also under control, with procedures laid down to monitor and control it.

Since the Bhopal Gas Tragedy involved the handling of risk from a hazardous substance, it inevitably led to a further tightening up of law in this regard. The important legislations passed are The Hazardous Waste (Management and Handling) Rules, 1989; the Manufacture, Storage and Import of Hazardous Chemical Rules, 1989; and the Public Liability Insurance Act, 1991. The last is to provide immediate relief to persons affected by accidents occurring while handling any hazardous substance. For expeditious disposal of cases arising out of such accidents and to provide early relief and compensation to persons, properties and environment, the National Environmental Tribunals Bill, 1992 has been introduced in the parliament. This envisages separate Tribunals to try environmental cases (*The Hindu Survey of the Environment 1993*, pp. 157–63).

Steps are also being initiated to introduce environmental audit of local municipal bodies, statutory bodies and public limited companies to evaluate the effect of their policies, operations and activities on the environment, particularly with respect to compliance with standards and the generation and recycling of wastes. Besides compelling the firms and other bodies to be conscious of the environmental implications of their actions, such annual statements are expected to help in identifying and focusing attention on areas of concern and initiating appropriate action. They will also be a source of information to the public (Government of India, February 1992, p. 11). The Policy Statement on Abatement of Pollution emphasises integration of environmental considerations into decision making at all levels. It mentions that to achieve this, the following steps are necessary:

- prevent pollution at source;
- encourage, develop and apply the best available practicable technical solutions;
- ensure that the polluter pays for the pollution and control arrangements;
- focus protection on heavily polluted areas and river stretches; and

- involve the public in decision making (Government of India, February 1992, p. 4).

India has also been actively participating in the global debate on the implementation of Agenda 21, adopted at the Rio Conference on Environment and Development in 1992. India has signed the Convention on Biodiversity and the UN Framework on Climate Change in 1992. Towards this end, India has strengthened its forestry programme and has reversed the earlier trend of decline in forest cover. Under an ambitious programme of afforestation, it has a target of covering over a million hectares each year, and recent figures show that these targets are actually being achieved. (See Annual Report of the Ministry of Environment and Forests, Government of India, for 1993–94, p. 66.) To counter deforestation due to pressure for biomass, India has launched an innovative programme of management of forests jointly with local people in selected areas since the early 1990s. This movement is catching up all over the country, under which local people participate in regenerating and managing forests and share their produce. India is also a signatory to the Montreal Protocol on Ozone Layer Protection along with its London amendment. Its provisions became effective for India since September 1992, and India has already started the process of phasing out ozone depleting substances.

Thus the evolution of environmental policy and its implementation have travelled a long way, particularly over the last two decades.

4.2 INSTITUTIONS AND INSTRUMENTS FOR INDUSTRIAL POLLUTION CONTROL

Direct regulation is the main instrument of pollution abatement in India, as is clear from the institutional background of legislation presented earlier. Though some of the institutions and instruments used in the control of industrial pollution have figured in the preceding section, they will be focused upon in this section to bring out their role and limitations. There are also two instruments in addition to direct regulation which are discussed in this section: citizens' suit provisions and economic incentives.

The apex policy making body in the field of environment is the Ministry of Environment and Forests, Government of India. It acts through the central and state pollution control boards for implementing anti-pollution laws. Set up previously to handle only water pollution under the Water (Prevention and Control of Pollution) Act 1974, the jurisdiction of these Boards was extended to cover air pollution too under the Air (Prevention and Control of Pollution) Act of 1981. The Boards have a two-tier administrative structure to carry out their functions under law. At the first tier, the Boards are constituted by the concerned governments (the central government in the cases of CPCB and respective state governments in the case of SPCB) by nominating two executive members—the Chairman and the Member Secretary, and 15 other members both official and non-official or honorary. These Boards are autonomous in nature and meet once in three months to guide the executive functionaries in their perspective, programmes and policies. At the second tier, the day-to-day activities of the Boards are carried out by the appointed regular staff (technical and administrative) of the Board under the guidance of the Chairman and the Member Secretary. While the Chairman of the Central Board (CPCB) is a full-time member, the Chairmen of State Boards (SPCB) may be full-time or part-time.

The Pollution Control Boards have no elected members. All members including non-official members are nominated by the central government in the case of CPCB and by respective state governments in the case of SPCBs. The CPCB has five or less official members representing the central government, five or less members representing the SPCBs, three or less non-official members representing agriculture, fishery, or industry or any other relevant sectors, and two members representing the companies or corporations owned and managed by the central government. Similarly, the SPCBs have five or less official members representing the state government, five or less members representing local governments like municipalities, three or less non-official members representing agriculture, fishery, industry or such other relevant interests, and two members representing companies or corporations owned by the respective state governments. There is no representation to voluntary agencies or NGOs, and environmental activists may or may not find representation. The constitution of CPCB, however, is such that the problems of the states are articulated through the representatives of SPCBs which also helps in proper coordination of policies.

The CPCB acts like the State Board in the case of union territories (which are under the central government) but may delegate powers and functions to a person or body stationed in the union territory. The CPCB also has zonal offices to cover the entire country so that they can be in closer touch with SPCBs under their control and monitor their functioning. The zonal offices are as follows:

1. Northern Zone Office, New Delhi, covering Jammu and Kashmir, Haryana, Himachal Pradesh and Punjab.
2. Eastern Zone Office, Calcutta, covering Bihar, Orissa, Sikkim, West Bengal, and Union Territory of Andaman and Nicobar Islands.
3. North Eastern Zone Office, Shillong, covering Arunachal Pradesh, Assam, Manipur, Meghalaya, Mizoram, Nagaland and Tripura.
4. Central Zone Office, Kanpur, covering Madhya Pradesh, Rajasthan, Uttar Pradesh, and Delhi.
5. Southern Zone Office, Bangalore, covering Andhra Pradesh, Goa, Karnataka, Kerala, Tamil Nadu, and Union Territories of Pondicherry and Lakshadweep.
6. Western Zone Office, Vadodara, covering Gujarat, Maharashtra, and Union Territory of Daman, Diu and Nagar Haveli.

In the same way, SPCBs have regional offices to cover the entire area of respective states. The Karnataka Board, for example, had till recently seven regional offices; this number has increased to 11 for a more effective operation.

The SPCBs are under a dual control of both CPCB and respective state governments. Their Chairman and Member Secretary are nominated by the respective state governments and can also be removed by them. The SPCBs are answerable also to the CPCB, and through it to the MEF, Government of India. There is a provision for joint Boards also between two or more states.

The function of coordinating SPCBs and also of laying down minimum national standards is carried out by the CPCB. The SPCBs are free to lay down more stringent, but not less stringent standards. The main function of carrying out the inspection of industries, monitoring their pollution level and implementing standards is done by the SPCBs. This is done by CPCB too in territories where SPCBs are not formed. The CPCB provides technical assistance to SPCBs to carry on and sponsor investigation and research on pollution problems. Both CPCB and SPCBs plan and organise training and awareness programmes on pollution prevention and control. They also collect, compile and publish technical and statistical data and prepare such materials which may be useful in pollution prevention activities.

The CPCB and SPCBs are also supposed to suggest economically viable methods to achieve the desired environmental quality. The Acts actually do not require the Boards to do so, but it is nevertheless supposed to be their function by convention. This puts the burden unnecessarily on the Boards, since the industries often take advantage of this by saying that there is no economically viable technology to achieve pollution standards recommended, instead of doing research and development in the field themselves or consulting other experts. If the pollution standards are not met, the industry should be closed down instead of continuing on the alibi that it is not viable to meet the standards. Due to severe resource constraints, the Boards cannot take on the task of developing pollution prevention and control technology for all conceivable situations. There should be incentive for the industry to develop and adopt it if prevalent technology is not satisfactory. Under the law and the pollution control mechanism evolved by it, the Boards have the necessary powers to enforce pollution standards and compel industries to adopt either clean technologies or clean-up technologies whichever the industries deem necessary, instead of allowing them to take cover under the alibi that technologies are not available to meet the standards.

The following regulatory instruments are used in pollution control or prevention mechanism: (*a*) No Objection Certificate (NOC); (*b*) Consent; and (*c*) Standards.

NOC is required in the case of new industries, and is essentially a certificate of site clearance to be obtained by the entrepreneur from the concerned SPCB for the proposed project. The concerned Board issues the NOC only after considering the impact of the proposed project on the specified environment, topographical and climate features, biological diversity in the area, compatibility with land use, proximity to protected areas (ancient monuments and archaeological sites, wildlife sanctuaries, national parks, etc.) and to human settlement.[2] This certificate is obtained once in the lifespan of an industrial establishment. Though a hazardous industry cannot be established near human settlements, there is little to prevent human settlements developing in the proximity of industry once it is established. There have to be adequate safety measures against hazards in any case.

Consent is required to be taken by an entrepreneur after the completion of the industrial project but before commissioning the industrial process. This is given subject to the installation of all required pollution control equipment to abate pollution. Without such a consent, the factory or industry cannot start its operation. The consent has to be renewed every year. It is supposed to be renewed only if pollution standards are complied with. The consent could be conditional or unconditional; or it may be refused or withdrawn in the case of inadequate pollution control measures (CPCB 1992, pp. 141–43). Starting or operating an industry without such consent is punishable with imprisonment of not less than two years and extendable up to six years in addition to a fine. While applying for consent, a fee is collected which is a source of funds for running the Boards. The consent fee can vary from state to state depending on the decision of SPCBs. In Karnataka, a single rate structure is applied based on fixed capital of the industry, both for air and water consent fees. The structure is as follows:

Fixed capital (Rs)	*Consent fee (Rs)*
More than 10 crore	15,000
5 to 10 crore	10,000
1 to 5 crore	7,500
50 lakhs to 1 crore	4,000
25 lakhs to 50 lakhs	2,000
10 lakhs to 25 lakhs	1,500
5 lakhs to 10 lakhs	1,000
1 lakh to 5 lakhs	250
1 lakh and below	100

1 crore = 10 million; 1 lakh = one hundred thousand.

Note: When land, buildings or open assets are taken on lease, twenty times the annual lease value is taken as capital investment.

Source: KSPCB (personally gathered, 1992).

In addition to the monitoring by SPCBs, the Health Departments of the Municipal Corporations are involved in monitoring industrial units to see if any of them cause health hazard to the public. Thus as per Karnataka Municipal Corporation Act 1976, all industrial units should obtain licence from the Health Department which is to be renewed annually.

Standards refer to specific parameters previously quantified with respect to measures for disposal, discharge and emission of solid, liquid and gaseous waste into the environment. These standards form the basis for the enforcement of any legislation for the abatement of pollution. Standards are based on the knowledge of pollutants, their chemistry and both short and long term effects on the receiving environment (Assocham 1991, p. 20).

In India, the Bureau of Indian Standards attended to the pollution problem as early as the 1960s and developed nearly 250 standards (IIEE 1991, p. 49). With rapid industrialisation, the need for industry specific standards was felt strongly in the early 1970s which were developed by mid-1970s. The main considerations in evolving them are: (*a*) available technology; (*b*) the cost of available technology; and (*c*) assimilating capacity of the environment. Since these standards are precise and representative, they have been adopted by the Government of India as 'Minimum National Standards' (MINAS) to abate pollution at the national level (Assocham 1991, p. 20). Though the standards were in existence even before the comprehensive environmental legislation, only in 1974 were they given statutory provisions for implementation. All the SPCBs are expected to adopt the MINAS and strive to implement them. While they have no powers to relax MINAS, they can make them more stringent depending on local conditions where ambient quality of air or water needs improvement.

Until recently standards were based on the concentration of pollutants in effluents and emissions. This provided scope for dilution which means wastage of water in addition to evasion of law. According to the notification (dated 19 May 1993) issued by the Ministry of Environment and Forests (MEF), Government of India, the following types of standards have been made mandatory from 1 January 1994. The standard types are:

1. General standards for discharge of environmental pollutants: effluent into
   (*a*) inland surface waters,
   (*b*) public sewers,
   (*c*) land for irrigation,
   (*d*) marine coastal areas;
2. waste water generation standards;
3. load based standards;

4. concentration based standards;
5. equipment based standards;
6. load/mass based standards;
7. process based standards;
8. noise standards.

The new category of mass based standards enables to set specific limits to encourage minimisation of waste, promote recycling and reuse, and more specifically conservation of resources.

The standards are not only a regulatory tool but also constitute a mechanism to promote technological upgradation to abate pollution and regulate waste (Government of India, June 1992, p. 7). There is a strong shift now from 'end-of-pipe' or 'clean-up' technologies to 'clean' technologies. While the Bureau of Indian Standards still looks after the technical aspects of standards, the actual responsibility of fixing standards for effluent discharge and gaseous emissions has been taken over by the MEF, Government of India. The Government of India has so far laid down industry-wise standards for 34 industries for both air and water, and noise and vehicular emissions (CPCB 1992). However, adoption and revision of standards is a dynamic process. In principle, standards are revised once in five years or so. This is due to the inclusion of new industrial types under the purview of pollution abatement and upgradation of available technology. But the revision of standards is also influenced by representations by industries and their associations (FICCI 1989, pp. 69–76; Assocham 1988, pp. 16–19; Assocham 1991, pp. 20–23).

The Ministry of Environment and Forests, Government of India, normally notifies four types of industrial effluent discharge standards. They are: (*a*) effluent discharged into inland surface waters, (*b*) effluent discharged into public sewers, (*c*) effluent discharged on land for irrigation and (*d*) effluent discharged into marine coastal areas. Accordingly, for the distillery units Biochemical Oxygen Demand (BOD) load of 30 mg/l was notified for all the four types. However, with the available technology, distilleries were unable to meet the notified standards and hence, a representation was made by the Distillers' Association to the MEF for the review of the notified standard. In response, the MEF re-fixed BOD value as 100 mg/l only for the effluent discharged on land for irrigation. Further, in case land application is envisaged as a secondary treatment system for further removal of BOD, then the BOD limit may be set at 500 mg/l or even raised up to 700 mg/l subject to regular and careful monitoring of run-off and leachate (letter dated 22 August 1989, MEF, Government of India).

The implementation of standards needs periodical inspection of industries. For this purpose, MEF has categorised industries into 'Red', 'Orange' and 'Green' according to the descending order of the level of the severity of pollution generated by the industries. The size of the industry is also taken into consideration for inspection—all large-scale industries are included in the 'Red' category. It should be noted that industries in the 'Green' category are not totally pollution free. Industries not involving pollution, and not requiring monitoring and inspection, are excluded from these three categories. The details of the frequency of inspection are as follows (see Table 4.1).

Table 4.1
Categorisation of Industries and Frequency of Inspection

Size of industry	*Category*	*Frequency of inspection*
Large	Red	Once a month, or once in 6 months depending on the severity of pollution
Medium	Red	Once in 3 months
	Orange	Once in 6 months
	Green	Once in a year
Small	Red	Once in 6 months
	Orange	Once in a year
	Green	Once in 2 years

Note: The size of the industry is as per fixed capital. The 'large' has a fixed capital of Rs 50 million and above; 'medium' has Rs 3.5 to 50 million; and 'small' up to Rs 3.5 million.

The Boards are vested with necessary powers to monitor and regulate pollution problems, and to obtain information for this purpose. On their own too, the Boards have the powers to conduct stream surveys to identify the flow of water and its characteristics. The Boards can give direction to any person abstracting water from or discharging effluent into any stream or well to give the required information about such abstraction or discharge. If it is not given, the persons concerned are punishable with imprisonment up to three months or with fines up to Rs 10,000 or with both. If non-compliance continues, the person is levied with an additional fine of Rs 5,000 per day till he complies with the direction. Besides, any Board officer can take a sample of water from an affected stream, well or sewage for investigation in recognised laboratories. Similarly, the Board officer can examine any record or other material objects in the premises of a factory or firm and conduct a search (CPCB 1992, pp. 17–20).

Persons causing pollution of air and water beyond the standards prescribed are punishable with imprisonment of not less than one and a half years and up to six years with a suitable fine. The Boards are empowered to take up the prescribed work in the firms seeking consent, and recover the expenses incurred in installing anti-pollution equipment from the persons or companies concerned. The enforcement mechanism of pollution abatement policy can be presented in a simplified way through Figure 4.1.

The foregoing discussion relates mainly to source-oriented standards. The Central Pollution Control Board also lays down effect-oriented or ambient quality standards for air and water, which are monitored by state level PCBs. These standards are tried to be met by stricter enforcement of source-oriented standards. Where water pollution is due to failure of municipalities and industries to treat water, bodies and mechanisms are evolved for coordinated action for an area, a conspicuous example of which is the Ganga Action Plan in northern India. Such coordinated plans are yet to become more widespread.

Shift of emphasis to administrative action

In spite of all the stringent provisions, the process of taking action was lax before the mid-1980s as the Boards had little scope for direct administrative action. They had to approach a magistrate to enjoin the polluter. The Amendments to the Air Act in 1987 and to the Water Act in 1988 not only strengthened penal provisions of these laws but also empowered the enforcement agencies to close down polluting industries and to stop their electricity and water supply.

> Earlier, few agencies had the financial and legal powers to speedily launch presentations and battle the cause through multiple appeals. With the change in the enforcement strategy,

Figure 4.1: The Pollution Abatement Policy Enforcement Mechanism

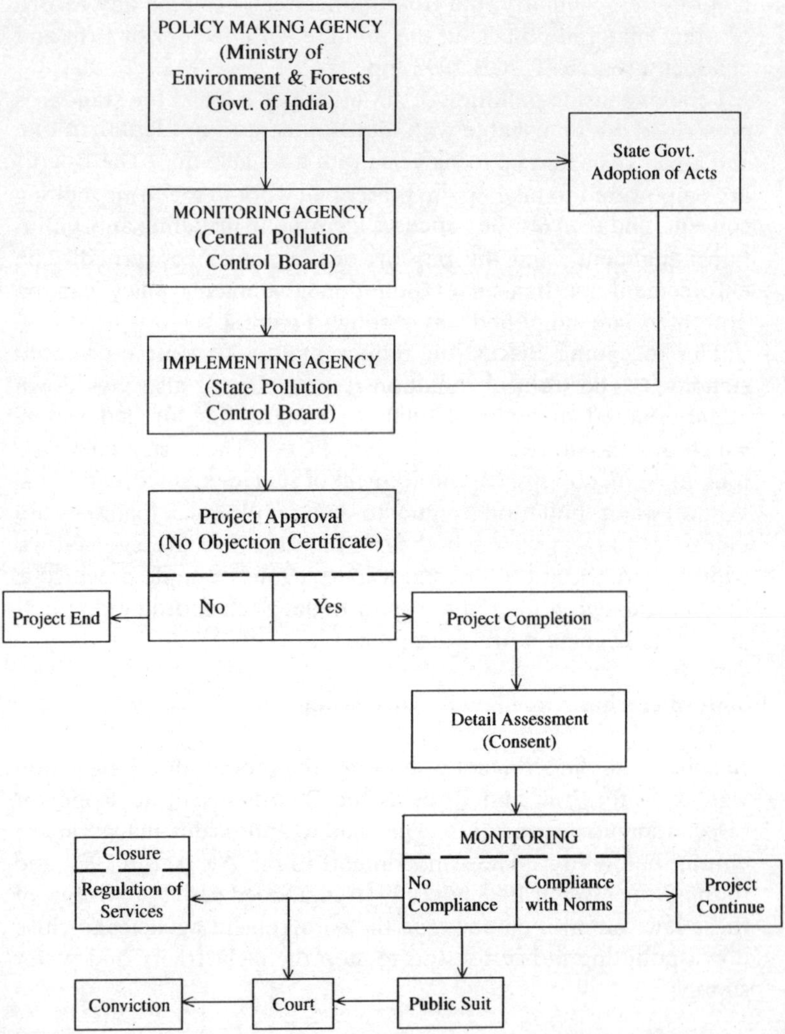

> however, the burden of initiating a court action now rests on an aggrieved polluter, who must challenge the agency's order in court. Consequently, from a lawyer's standpoint, the agency merely has the task of defending its administrative order (Rosencranz et al. 1991, p. 2).

This shift from judicial to administrative enforcement of environment laws is meant for better compliance. However, the Boards still have no powers to exact fines, order imprisonment or compel compliance with their detections, and have to depend on the courts of law.

Citizens' suit provision and role of judiciary

Environment is too important a matter to be left only to government agencies. If the agencies, for various reasons, do not take proper action against the polluter, the citizen has a judicial choice in India in the form of three civil remedies: (*a*) a common law *tort* action against the polluter; (*b*) a writ petition to compel the agency to enforce the law; and (*c*) a citizens' suit to enforce statutory compliance (Rosencranz et al. 1991, p. 77). It may be recalled that with the introduction of Article 51–Ag into the Constitution by the Constitution (42nd Amendment) Act 1976, the citizens were also made responsible for environmental protection. But the citizens had no power then to prosecute a polluter. Section 19 of the Environment (Protection) Act of 1986 and subsequent amendments to Water and Air Acts brought about a change in this provision. According to the new regime, a citizen may prosecute an offender by a complaint to a magistrate. However, to enable the government to take appropriate action against a polluter, the citizen must give 60 days' intimation notice to the government (Rosencranz et al. 1991, p. 118).

With the emergence of Public Interest Litigation (PIL) (whereby even citizens not directly affected by pollution can launch litigation in public interest), the role of the higher judiciary (the High Courts and Supreme Court) has undergone a radical and creative transformation. In the process, the Supreme Court has interpreted the right to life and personal liberty to include the right to a wholesome environment. The Supreme Court embarked even on necessary administrative action such as appointing expert committees, and based on their findings, giving direction for closure of polluting industries. The closure of tanneries polluting the river Ganga and directing municipalities to prevent wastes from flowing into the river is an illustration of how the Supreme Court played an active role. In the Limestone Quarrying case, the Supreme Court ordered the closure of the quarries and even supervised the regeneration of the devastated Dehradun valley. The High Courts have also taken similar action, as for example, when the Andhra Pradesh High Court did not allow the Life Insurance Corporation and Income Tax Department to build residential houses in a recreational zone (Rosencranz et al. 1991, pp. 2, 3, 57 and 58).

When citizens are empowered to prosecute a polluter through PIL, they need authentic information to fight their case. There is the right to information from government records, that is, public access to government information has now been granted by Supreme Court decisions. This right to information has been derived by the judges from the fundamental right to freedom of speech and expression, and the fundamental right to life and personal liberty (Rosencranz et al. 1991, p. 143).

With an increasing number of cases based on environmental pollution and destruction coming up before the courts, the Supreme Court had suggested as early as in 1987 to set up separate Environmental Courts. The cases need special expertise, which a separate network of courts could well afford to organise. This proposal is yet to be implemented. Section 4.3 assesses the functioning and effectiveness of the various types of direct regulation and juridical approaches for pollution control.

Apart from industry specific standards, Central Pollution Control Board (CPCB) has developed ambient water and air quality standards. There is a regular monitoring of ambient water and air quality throughout the country. Considering the size of the country, the number of monitoring stations seems to be inadequate, particularly for air. For example, there are only eight stations for monitoring air quality in Karnataka while there are 72 for monitoring water quality.

Economic incentives

Complementing the measures for direct regulation and judicial processes, there are some economic incentives to induce firms to implement pollution control measures and desist from wastage of water.

We have already referred to the *water cess* in Section 4.1. As per the Water (Prevention and Control of Pollution) Cess Act of 1977, persons and local authorities carrying on industrial activities are levied a cess on the quantity of water consumed at rates mentioned in Table 4.2. The maximum rates as applicable since the Act came into force were doubled in 1992, and were made still higher for those not complying with the Water Pollution Control Act of 1974. There is a further rebate of 25 per cent for implementing pollution control measures and meeting the standards, over the lower rates applicable to those complying with the Act.

Table 4.2
Water Cess Levied on Water Consumption by Type of Use (as revised in January 1992)

Purpose	*Maximum rate for those complying with the Act*	*Maximum rate for those not complying with the Act*
	Rates per kilolitre	
Industrial cooling, spraying in mine pits or boiler feeds	One and a half paise	Two and one-fourth paise
Domestic purpose	Two paise	Three paise
Processing whereby water gets polluted and pollutants are easily bio-degradable	Four paise	Seven and half paise
Processing whereby water gets polluted and pollutants are not easily bio-degradable	Five paise	Nine and half paise

Note: One paisa is one-hundredth of a rupee (paise is plural of paisa).
Source: CPCB 1992, p. 107.

Though the primary purpose of the cess was to provide some funds to the PCBs, it is also expected to induce saving of water. It is not, however, as effective a deterrent to pollution as in the case of water effluent charges in the Netherlands because its incidence in India is low compared to the cost of pollution control.

As has been argued earlier, the water cess hardly meets the purpose of an effective charge on pollutants. The incidence of this cess is quite low and cannot serve as an effective incentive for pollution control. For the industrial units taken up as case studies by us in India (see Chapter 6), Table 4.3 indicates the magnitude of water cess paid and its proportion to annualised costs on pollution control and total annualised costs of the respective industrial units.

Table 4.3
Water Cess Paid by Industrial Units in India
(Average of 1989–90 to 1991–92)

Industrial unit	*Water cess paid in Rs*	*As per cent of annualised costs on pollution control*	*As per cent of total annualised costs*
Textile industry			
Minerva	4,986	1.08	0.0152
Srinivasa	5,494	1.52	0.1157
Gemini	8,966	0.62	0.0280
Binny	18,485	5.13	0.0569
Gogte	1,433	0.43	0.0040
DCM	5,408	1.26	0.0140
Average	7,462	1.41	0.0194
Cement industry			
Rajashree	22,881	0.14	0.0036
Mysore	13,519	0.12	0.0031
Kanoria	10,000	0.17	0.0073
Average	15,467	0.14	0.0038
Fertiliser industry			
MCF	47,006	0.15	0.0072

Source: Questionnaires canvassed for respective units.

Depreciation Allowance for industries on devices and systems installed for reducing pollution and for conservation of natural resources is a positive inducement. A list of devices and systems that are permitted such an allowance is notified by the government (Government of India, Extraordinary Gazette, Part II, Section 3, Subsection ii, No. 85 dated 23 February 1983, Annexure II). When first introduced in the 1982–83 Budget, the depreciation allowance was only 30 per cent, but was raised to 50 per cent in the 1988–89 Budget. The 1991–92 Budget reduced it to 40 per cent, while the 1993–94 Budget raised it to a full 100 per cent—a position which has continued to date. An extra depreciation allowance up to a maximum of an amount equal to one-half of the normal allowance is allowed if the concern works on two shifts, and normal allowance is allowed for working three shifts. The depreciation allowance is allowed for installation, and not necessarily for actually operating the equipment—let alone operating it for all the shifts worked. It is assumed that the devices installed are worked, since other provisions of the law are expected to ensure that no pollution is let out into the environment beyond standard limits.

In addition to depreciation allowance, there has also been an *investment allowance* since 1983–84 at the rate of 35 per cent of the actual cost of new machinery or plant which would assist in the control of pollution or protection of the environment. This is allowed on notified plants and machinery (Government of India 1988).

There are also *exemptions in indirect taxes* so that the burden of cost of pollution control is reduced. Equipments manufactured for pollution control have been exempted from central excise duties since 1990. Concessional customs duty is levied on notified equipments and spares imported for pollution control. There is concessional excise duty on goods manufactured from industrial waste, for example, there is 100 per cent excise relief on paper manufactured with 75 per cent bagasse (sugarcane residue). Bricks manufactured using more than 50 per cent fly ash are completely exempt from excise (Assocham 1991, p. 27). In Karnataka, pollution control equipments have been exempted from sales tax since April 1991.

There is an *exemption from income tax* on amounts contributed for programmes of conservation of natural resources. These amounts are deductible from taxable income. Such a step is aimed at involving the general public, particularly, the corporate sector and individuals in environmental protection, and was introduced with effect from June 1982 (Government of India 1988).

There is a tax concession for *decongestion of industries*. To encourage shifting of industries from overcrowded cities and to reduce pollution, the capital gains arising from transfer of buildings or lands used for the purpose are exempt from tax if these are used for acquiring land or constructing buildings for the purpose of business in a new place. Since 1983–84, this exemption has been extended to capital gains arising from transfer of machinery and plants (Government of India 1988).

An *Environment Relief Fund* was created by the Public Liability Insurance Act 1991 to provide relief to any person affected while handling any hazardous substance. This is an insurance policy taken by an undertaking which deals with hazardous substances. The value of the said policy is not less than the paid-up capital of the undertaking. The Insurance Company ensures that proper precautions are instituted in handling hazardous substances so as to reduce its risks for which it provides cover.

Soft Loans are another form of economic incentives to encourage installation of pollution control devices and systems. They carry a lower rate of interest of 11.5 per cent per annum as against the normal lending rate of 14 per cent for upgradation of technology/ modernisation of existing facilities. These soft loans are given by all-India financial institutions which normally extend credit to industries. The Industrial Finance Corporation of India (IFCI) has started a new scheme called 'Scheme of Subsidy for Control of Pollution in Small and Medium Scale Industrial Units' to meet the growing need for formulation of suitable proposals for prevention and control of pollution in industrial units, which meets the approval of the concerned authorities like SPCBs. This scheme envisages subsidising the fee payable to the Technical Consultancy Organisation for preparing, and if so required, implementing a scheme of pollution control in specific industrial units to the satisfaction of the regulatory bodies or SPCB (Assocham 1991, p. 27).

Unfortunately, the soft loan scheme is not as extensive as it should have been. For example, the commercial banks have no such instruction to give soft loans to industries to install pollution control infrastructure. The banks, however, normally insist on the No-Objection Certificate from the concerned PCBs when extending loans to industries. An industry is not likely to receive any loan for installing pollution control equipment separately, if it is a loss making unit and if it is not in a position to return the loan.[3] The industry can however get bank loans at normal rates of interest for installing pollution control infrastructure if it is or expected to be economically viable and is able to return the loan, irrespective of whether such installation is part of an overall package for the industry or is just an add-on, separate measure.

4.3 EFFECTIVENESS OF PAST AND PRESENT POLICIES

We shall first note the things that have contributed to making the environmental policy more effective, particularly in recent years.

We shall then note other things that have tended to make it less effective. This would indicate the scope for improvements and corrections in policy so as to make it more effective.

4.3.1 Positive developments

The environment policy and its working which was piecemeal and compartmentalised in the past particularly up to the 1970s has become more integrated and holistic since the 1980s. We have traced the evolution of the policy in Section 4.1. It is very interesting that (as will be seen in Chapter 5) a similar process took place in the Netherlands too and what is more, at about the same time periods. A proper coordination was made possible with the MEF at the apex and the PCBs acting as its implementing agencies for dealing with industrial pollution. An integration of both air and water pollution control is important especially due to the possibility of transfer of pollution from one medium to another. The Boards could have an integrated approach as they deal with both. On the judicial front too, special Environmental Tribunals are proposed to deal with cases involving hazardous substances, if not all environment related cases. This is for quick redressal of grievances of affected parties. The emergence of proper institutions for the implementation of environmental policy is one of the most heartening developments contributing to its effective implementation.

The approach to the implementation of environmental policy has also undergone a few welcome changes. Environment legislation before the 1970s allowed only tort action under common law by the affected victims. In a developing country where the victims of environmental degradation are poor and often unorganised, such an approach could not guarantee against pollution. It needed a lot of time, patience and money on the part of the affected victims, who also had the burden of proof lying with them. With the passing of Water Act 1974 and Air Act 1981 and with the creation of PCBs, these institutions, in addition to the affected victims, could book the offenders. Furthermore, with the Environment Protection Act 1986 and amendments to Water and Air Acts, emphasis shifted from a judicial approach to administrative action since the Boards were empowered to close down offending industries and to stop the supply of electricity and water to them. The burden of proof of innocence shifted to industries. This altered the scope for implementation for the better, making the Boards more effective than in the past.

With the citizens' suit provision and with the right to information with the government (and PCBs) guaranteed under such provisions, even third parties could launch PIL against offenders. This made the PCBs more alert and reduced the possibility of secret deals between Board officials and offending industries. These provisions have greatly helped ecological movements, environmental activists and NGOs, making them a more potent force that can hardly be ignored by the polluting industries. Ecological movements have become so bold that they do not hesitate to take extra-legal actions where they feel that Board officials are deliberately avoiding the taking of proper action against offending industries. There are instances where regional offices of PCBs have been attacked and ransacked. Paradoxically, power to initiate legal action under PIL and the tendency to take extra-legal action have grown simultaneously. Both have indicated that neither the Boards nor the polluting industries can take people for granted. The result is that industries, particularly those of the large and medium scale, have begun to feel that anti-pollution measures are unavoidable and they must somehow implement them. Following PIL, the higher judiciary (Supreme Court and High Courts) have also been playing an increasingly active role in the cause of the environment (discussed in the preceding section). This has increased the riskiness of ignoring environmental law for industries, in the form of delay in the execution of industrial projects and cost over-runs.

The shifting of environment policy to the Concurrent List and assumption of greater powers by the central government have also helped in evolving a national policy and contributed towards more or less uniform implementation. This was necessary because otherwise, there could be a tendency among the state governments in a federal set-up to compete among themselves for attracting more industries and investments in their respective states by relaxing environment and labour laws. However, the states still have a say, and the major task of actually implementing the policies is left to them. Nevertheless, the picture is much better now than in the early 1970s, and the states cannot disregard national policies and institutions which they have themselves helped in evolving.

Apart from legal compulsions, it is important to make the industries increasingly conscious of the environmental problems they create. In this direction, two proposals mooted by MEF, Government of India, are of interest. One is the amendment of Company Law to make the publication of 'Environmental Audit' mandatory for industrial firms along with the financial audit. This could reduce the monitoring burden of the PCBs and help them take speedier action. The compulsion to publish environment audit and the knowledge that it could be checked, makes the industries more responsible. The second proposal is to create the post of a Chief Environmental Officer in each large-scale and medium-scale industry who can monitor energy efficiency and pollution control in respective industries. It is applicable to both existing and new industries. This need not necessarily entail additional expenditure as existing officers in the industry are allowed to take additional charge of the environment. The MEF has also instituted annual awards to industries for best 'environment-friendly performance'.

Another heartening sign is the increasing awareness on the part of both industry and government to shift from end-of-pipe treatment or clean-up technologies to clean technologies. Both the policy statements (Government of India, February 1992, June 1992) have voiced the need for such a shift. It is realised that the clean-up technologies merely transfer pollution from one medium to another and may not enable an overall view, apart from being costlier. However, we have still a long way to go before clean technologies are adopted on a significant scale. In any case, only new industries and those which intend to modernise their plant can go in for it; not the existing ones. There is a possibility, however, that in the long run (hopefully not too long) such a shift is going to take place. But even in the short run, the net result of making environmental law more stringent, building up institutions for its monitoring and implementation, increasing incidence of PIL and a more active role of the high judiciary, have all combined to compel industries to take into account environmental considerations in their project design and decision-making process.

This awareness of unavoidableness of pollution prevention and control on the part of industries has led to some new institutional developments. A few companies specialised in pollution control technology have emerged which are hired by polluting industries to install and commission pollution control equipment on turnkey basis. This has meant greater professionalisation and efficiency in the area. Another institutional development is the emergence of common treatment plants particularly by small industries who cannot afford to have their own separate facilities. The PCBs help in promoting such plants. Half of the cost of pollution treatment is shared on mutually agreeable principles and the other half is given as subsidy by the central government. There is one such plant for tanneries in Bangalore in Kadugondanahally. Thus the financially weak small units are also brought within the ambit of pollution control mechanism.

4.3.2 Shortfalls and constraints

Though policy and legislation may look stringent on paper, there is many a slip between the cup and the lip. The implementation of pollution control in the 1990s is certainly much more effective than it used to be in the 1980s and far more so compared to the situation during the 1970s. Yet, there are shortfalls and constraints reducing the effectiveness of environment policy regarding pollution control. There is thus significant scope for improvement. It would be useful in this connection to have some indication in terms of hard figures about how far the pollution control law is put into effect. One such indication is how many industrial units have at least set up treatment plants. This can be seen from Table 4.4 for two recent years. Unfortunately comparable figures are not available for previous years. But as per the judgement of Karnataka State Pollution Control Board (KSPCB) officials, the situation has improved considerably over the years. Even among the units where pollution treatment plants were not in operation, there were a significant number which had started installing the plants or planned to install in the near future.

There is a conspicuous increase in the proportion of units with air pollution treatment in operation between 1991 and 1994. Air pollution control lagged behind till 1993 and even now there is a greater laxity in this case. This is possible because of the feeling that in a developing country while the control and prevention of water pollution is crucial, it is not so in the case of air. There may be a serious concern about preventing hazardous substances leaking into the air, especially after the Bhopal Gas Tragedy, but the

Table 4.4
Proportion (%) of Industrial Units with Pollution Treatment Plants in Operation in Karnataka

	As on 31.3.91	*As on 31.3.94*
For water pollution control		
Large industries	65.8	72.3
Medium industries	65.0	67.1
Small industries	NA	46.3
For air pollution control		
Large industries	33.8	73.3
Medium industries	33.5	76.4
Small industries	NA	65.9

Source: Annual Reports of KSPCB, 1990–91 (pp. 8–10) and 1993–94 (p. 9).

release of slow poisons into the air and long-run effects of greenhouse gases are treated with relative complacency. This is because the cost of compliance is high and it is feared that the industries are not able to absorb these costs without adversely affecting growth rates of output and employment. It is not clear whether the sudden improvement in air pollution control is unique to Karnataka or is true for the country as a whole.

Though Table 4.4 relates to treatment plants actually in operation, it is possible that they may be kept in operating condition or operated only part of the time but not throughout the working of the industry. This is to save energy, since energy costs are high and keep increasing. Moreover, even if industries are prepared to pay for energy, the supply of energy falls short of demand and the gap is increasing. Though separate supply of energy is ensured to run pollution control plants, this energy is often stealthily used to run the main factory itself. Though the authorities try to minimise this by installing separate meters to measure the consumption of electricity for pollution control plants and for normal production process, the industries have their own ingenious way of overcoming such problems. The tendency of not operating pollution control plants even after their installation for the purpose is in evidence both in the case of air and water pollution but particularly so in the case of air. This points to the need for adopting process-integrated technologies for prevention and control of pollution instead of add-on technologies, at least wherever possible.

Even highly polluting industries do not all comply with standards, though the situation has improved over the years. As per the Annual Reports of KSPCB for 1992–93 (1993, p. 45) and 1993–94 (1994, pp. 46–47), of the 99 industrial units in the highly polluting category, only 38 had adequate facilities to comply with the standards in 1992–93; and only 46 of the 85 units in 1993–94. Though this proportion shows an improvement from 38 per cent to 54 per cent within a year, it also shows that a significant number of industrial units do not comply with standards even in a category where implementation should have been strict. Karnataka lags behind the country as a whole in this respect. As per CPCB Annual Report for 1993–94, 74 per cent of industrial units in the highly polluting 17 categories of industries in the country had adequate facilities for compliance with pollution control as against only 58 per cent in the state.

What is the regulation machinery doing about it then? As per the Annual Report of KSPCB (for 1993–94), only 87 prosecutions were launched under the Water Pollution Act of 1974 and 37 under the Air Pollution Act of 1981 since the inception of the Board till 31 March 1994. These prosecutions cover the cases of both highly polluting and other industries. The number of industries ordered to be closed down by the Board for non-compliance was much fewer—only 12 under the Water Act and 10 under the Air Act. The powers of the Board have been rather sparingly used.

Public Interest Litigation is an important and fairly active provision which keeps the Board and industries alert. In such cases, both the PCB and the non-complying industries are made defendants. As many as 65 PIL cases were filed as on 31 March 1994. The position in this regard can be seen from Table 4.5.

A strict implementation of pollution control regulation should get reflected in the ambient air and water quality. The Annual Reports of CPCB and KSPCB give no indication of trends in ambient air quality, but there is some information about ambient water quality of major rivers in the country (see CPCB Annual Report for 1992–93). An extract of this information for Karnataka is presented in Table 4.6 which shows a deterioration in water quality at all locations in regard to total coliform, and either no trend or upward trend for dissolved oxygen, and a mixed picture for biochemical oxygen demand. This is hardly a rosy picture.

Table 4.5
Prosecutions Launched and Convictions Secured under Pollution Control Laws (as on 31 March 1994)

	Water Act	*Air Act*	*Writ in High Court*	*PIL*
No. of prosecutions launched (complaints in the court)	87	37	51*	65
a) No. of cases decided	44	14	58**	24
i) decision in favour of the Board	20	3	46	22
ii) against the Board	24	11	12	2
b) Pending	66	23	93	41

* during the year
** including previous cases
Note: PIL: Public Interest Litigation.
Source: Karnataka State Pollution Control Board (KSPCB) Annual Report, 1993–94.

Table 4.6
Trends in Water Quality of Rivers at Selected Locations in Karnataka State

River/ Location	*Period*		*Dissolved oxygen*	*Bio-chemical oxygen demand*	*Total coliform*
	From	*To*			
1. Krishna at Tintini Bridge	1979	1991	NT	DN	UP
2. Tungabhadra at Ullanur	1979	1991	NT	DN	UP
3. Bhadra at Malleswaram D/s of Kiocl	1979	1991	NT	DN	UP
4. Cauvery at KRS Dam, Balamurik-shetra	1980	1991	UP	NT	UP
5. Kabbani	1987	1991	NT	UP	UP

NT: No trend; UP: Upward trend; DN: Downward trend; D/s: Downstream.
Source: Annual report of Central Pollution Control Board for 1992–93, pp. 24–26 (extracted from All-India Table).

Hazardous wastes continue to be a grave problem in the country as well as in the state. Through a study assigned to a consultant firm by Karnataka SPCB, it was gathered that the total generation of waste in the state was as follows (see KSPCB Annual Report 1993–94, p. 15):

Solid	390,403	Tonnes/year
Semi-solid	368,770	Tonnes/year
Liquid	29,273,183	Tonnes/year

It was also found that Bangalore district alone accounts for almost 50 per cent of the total hazardous waste generated in the state. The Board has decided to identify waste disposal sites in respective districts, conduct Environmental Impact Assessment (EIA) for each of them, and examine the possibility of recovery and recycling so that the final quantity to be disposed at the site is minimum. (See KSPCB Annual Report 1993–94, p. 16.)

The implementation of environmental laws has to be monitored effectively at the industry level by the PCBs if the environment policy is to be effective. There are, however, a few important impediments in their way. The very first is the inadequate budget and staff of the Boards which is particularly true of SPCBs: a case illustrated by the KSPCB. No doubt, its total revenue increased from a mere Rs 2.186 million in 1981–82 to Rs 26.952 million in 1991–92 and to Rs 41.416 million in 1993–94. (Adjusting it for the increase in the general price level, i.e., in index numbers of wholesale prices for India, these figures in terms of 1981–82 prices work out to Rs 11.979 million in 1991–92 and Rs 15.991 million in 1993–94). On the other hand, there were 138,000 industrial units in the state in 1993 as per the Department of Industries, inclusive of small-scale units. This means there were only Rs 300 (at current prices) available to monitor one industrial unit in 1993. A welcome change in the budget position is the increasing reliance on revenues generated by KSPCB itself, such as water cess, consent fees and fines, and decreased reliance on grants from the governments (state and central). This is consistent with the autonomous status of the Boards. The composition of the source of finance of KSPCB is shown in Table 4.7.

Since water cess is collected under a central government Act, SPCBs are the collecting agencies and the proceeds are first credited

Table 4.7
Sources of Finance of Karnataka State Pollution Control Board (KSPCB)
Rs in Million (percentages to total in brackets)

Sources	*1981–82*	*1991–92*	*1992–93*	*1993–94*
1. Grants from government and CPCB	1.720	2.738	5.339	3.706
	(78.7)	(10.2)	(13.4)	(9.0)
2. Water cess reimbursement	–	3.597	4.095	3.776
		(13.3)	(10.3)	(9.1)
3. Fees	0.017	9.567	19.195	23.279
	(0.78)	(35.5)	(48.1)	(56.2)
4. Others	0.449	11.050	11.296	10.665
	(20.5)	(41.0)	(28.3)	(25.7)
Total	2.186	26.952	39.925	41.416
	(100)	(100)	(100)	(100)

Note: 'Others' include own resources of KSPCB such as interest, fines and carry over cash.
Source: KSPCB Annual Reports.

to CPCB, which are thereafter reimbursed. Such reimbursements are shown against Item 2 in Table 4.7. Other assistance received from CPCB is included in Item 1. The procedures of first crediting the cess proceeds to CPCB followed by reimbursement are avoidable delays. A way could be found to simplify matters so that SPCBs can directly retain the cess collected under authorisation by the CPCB.

The inadequacy of the staff strength of the SPCBs can be brought out again with the example of Karnataka SPCB. It has a staff of about 205 including both technical and administrative as against the sanctioned strength of 358. Its jurisdiction covers the entire state of Karnataka whose geographical area is 191,791 sq.km. As against 138,000 industrial units in Karnataka at the end of 1993, the number of industries identified for pollution monitoring by the Board was only 8,966 during 1992–93, which is only 6.5 per cent of the total. Even this is a fairly large number to do justice to, compared to the available financial and manpower resources. The dearth of trained manpower in the country is not the main constraint, but it is the inadequate allocation of funds to SPCB which comes in the way of adequate staffing.

Even the existing staff has to face impediments in functioning. The Boards somehow assume the burden of indicating what technology to adopt to reduce pollution, if the industries are unable to cope with pollution. Such a technology has also to be 'economically viable'. A deterrent charge on effluent and emissions would force the firms to seek a proper technology from consultants in the field and adopt it. If the Board is unable to indicate an economically viable technology to treat pollution, then the firm feels obliged to be indifferent to the environmental degradation that goes on. The mood in the country is still not ripe for shifting the emphasis from economically viable technology to best available technology. In the absence of a deterrent charge on effluent, there is also no incentive to evolve and adopt such a technology.

There are some shortfalls in the development of pollution control technology, though pollution control equipment is easily available and the technology for their manufacture is established in India. There are over a hundred indigenous manufacturers of pollution control equipments including waste water treatment plants.

> A number of these manufacturers have well established fabrication departments but generally lack the capacity and interest to improve upon the process technology related to local or imported technology so as to make their equipment amenable to local conditions. A buyer of pollution control equipment should therefore make it a point to ensure that the manufacturer has his own design department and that the manufacturer supplies equipment as per customers' need and not as per the design that the manufacturer may have obtained from other sources (Shah 1992).

Most of the present standards are based on concentration of pollutants in effluent and emissions. The standards can be met simply by diluting the concentration with the help of more water. However, by resorting to such a method not only are the mass of pollutants not reduced, it also leads to wastage of water. A shift to mass based standards is now mooted to solve this circumvention of standards. It may take some time to shift to the new regime of standards.

The most glaring lacuna is in the case of small-scale industries because of their large numbers. The requirement of consent to be obtained from the Boards by industries and its annual renewal is implemented mainly in the case of large- and medium-scale industries and in the case of small-scale industries under 'Red' and 'Orange' categories. Small-scale industries under the 'Green' category are too numerous to be monitored properly. This is to some extent compensated by monitoring conducted by health inspectors of municipalities. In actual practice, there is considerable relaxation by the authorities concerned in the case of small-scale industries, as it is often uneconomical for these industries to install their own treatment plants. The Board however tries to set up common collective treatment plants for small industries and recover some of the costs from the industries. This requires that the concerned small industries are located together, as is usually the case. Where units in the same type of industry are not located together, they generally let out effluent in municipal sewage, leaving the municipalities to tackle the problem. On the whole, however, the pollution problem created by them is very inadequately tackled if not totally ignored, and the priority of the PCBs is mainly to deal with the pollution caused by large and medium industries.

Another obvious flaw is one seen in checking vehicle emissions. Though the concerned authorities try to check cars and two-wheelers, that too occasionally, the bigger offenders like public transport buses, lorries and autorickshaws are left free. The transport vehicles are not yet designed to reduce air pollution and unleaded petrol has only recently been introduced in the country. There is a similar disregard for pollution at the consuming end in other respects too. For example, several drugs and pesticides like DDT banned in other countries are allowed to be manufactured in India. The use of low quality detergents polluting water continues to add to the problem of treatment of sewerage.

There are thus conflicting influences at work. While on the one hand, environmental consciousness is increasing leading to tightening up of environmental laws, there is at the same time evidence of resistance to them and laxity in implementation. On the whole, the situation is better today than it was a decade or two earlier, but we have still a long way to go. The main source of 'consolation', if we may call it such, is that due to the low levels of income, demands on environment are low, even after allowing for a large population. Though India has a much larger population, the total emissions from fossil fuels and cement manufacturing amounted to only 178 million tonnes of carbon in 1989 (0.21 tonne per capita), compared to 1,329 million tonnes in USA (5.34 tonnes per capita). On the other hand, in generating each million dollar of gross domestic product (GDP) in 1989, India emitted 670 tonnes of carbon dioxide while USA emitted 259 tonnes and high income countries only 186 tonnes (World Bank 1992, p. 204). On deeper reflection, we cannot be complacent about the fact that the total demand on the environment from low income countries like India is low. If they are launched on a high growth path while persisting with present technologies of using the environment, it may have disastrous consequences. It is a matter of urgency that such countries are helped in shifting over to cleaner technologies so that higher growth can be attained at low environmental cost.

4.4 COSTS OF ENFORCEMENT AND COMPLIANCE

Is the cost of enforcement and compliance of pollution control laws a deterrent in their effective implementation? If this is so, strong resistance to enforcement is understandable. Fortunately, this does not appear to be so in the bulk of the cases.

The cost of implementation of pollution control measures has two components: (*a*) the cost of enforcement borne by agencies such as the PCBs; and (*b*) the cost incurred by the industrial units or persons towards adopting pollution control measures.

Before we proceed to present some estimates of these costs, it would be apt to keep in mind some limitations of these estimates so that we can be cautious in interpretation. If the funds at the disposal of an enforcement agency are inadequate for an effective enforcement, the estimates of cost of whatever enforcement is done would be lower than what they really would have been in the case of effective implementation. Similar difficulty arises in the case of industries too. If in the case of effective implementation they do not operate the pollution control plant, its operation costs would be limited. In such cases, it is better to take only firms which implement pollution control measures, and the costs which are needed to be incurred for compliance rather than actual costs. Another source of underestimation arises where clean technologies are adopted instead of clean-up technologies. Costs of controlling pollution after it arises are explicit. But if it is prevented at source, an increase in the cost incurred by adopting such a clean technology is not explicit. Fortunately, there is evidence to the effect that it is often much cheaper to build-in pollution prevention and control in the design stage than to adopt it as an add-on measure (Royston 1979, Chapter 7). The costs of compliance discussed here are costs of clean-up technologies.

4.4.1 Cost of enforcement

Strictly speaking, in estimating these costs, expenditures incurred by the MEF, CPCB and SPCBs have all to be included. If we consider the costs at the state level, such as Karnataka, the costs incurred by MEF and CPCB have to be appropriately apportioned. In the absence of full details of these costs, we have considered here the costs incurred by the Karnataka SPCB, since it nevertheless gives an idea and after all it is the main implementing agency. These costs are incurred on inspection, taking samples, laboratory tests, legal proceedings, etc. The main items of expenditure are salaries, laboratory equipment and chemical inputs, and other overheads. The average cost of implementation of pollution control measures by the Board per unit of industry varies with different categories of units according to the time and other resources devoted to monitoring them. Based on the discussion with the SPCB staff, it is gathered that an industrial unit in 'Orange' and 'Green' categories requires respectively one-third and one-sixth of the total resources required for an industrial unit in the 'Red' category.

In Karnataka, the SPCB spent Rs 57.25 million on 7,715 industrial units during 1991–92 and Rs 66.50 million on 8,966 industrial units during 1992–93. The average cost of enforcement per industrial unit turned out to be Rs 7,417 during both the years. During the two years 1991–92 and 1992–93 respectively, the SPCB covered 2,316 and 2,672 industrial units in the 'Red' category, 1,698 and 1,992 industrial units in the 'Orange' category, and 3,705 and 4,302 units in the 'Green' category. The proportion of these three categories to total industrial units covered was practically the same during both these years, namely, 30 per cent, 22 per cent and 48 per cent respectively.[4] Though the 'Green' category is preponderant numerically, the 'Red' category is preponderant in terms of time and expenditure devoted. Taking into account the weights for these three categories (1: 1/3: 1/6), the expenditure on enforcement incurred by the Karnataka SPCB on the three categories is presented in Table 4.8.

Table 4.8
Cost of Enforcement Incurred by KSPCB—Category-wise

	Red	*Orange*	*Green*	*All industries*
Per unit expenditure (Rs)				
1991–92	16,359	5,453	2,727	7,417
1992–93	16,408	5,469	2,735	7,417
Total expenditure (in million Rs)				
1991–92	37.9	9.3	10.0	57.2
1992–93	43.8	10.9	11.8	66.5

Source: KSPCB, Bangalore (category-wise figures are derived from weights).

Surprisingly, the 'Green' category receives more attention than the 'Orange' in terms of total expenditure. A shift of emphasis to 'Orange' between these two categories may be more in order. A more important point, however, is the level of enforcement costs, which appear to be only marginal in relation to the turnover and investment of firms, and are recoverable from them without adversely affecting their growth and profitability. The enforcement costs are low also in relation to pollution control expenditure incurred by the firms.

4.4.2 Cost of pollution control incurred by firms

CPCB brings out Comprehensive Industry Documents for selected industries occasionally, outlining the status of pollution control in respective industries and costs incurred by them in controlling pollution. These documents derive the annualised burden of costs by taking into account depreciation of capital (at 10 per cent linear rate), interest, operation and maintenance costs, and express them as percentage of annual turnover. These documents are not brought out every year for each industry and hence trends and costs cannot be surmised. Nevertheless, they enable us to have an idea of the cost burden.

The annualised cost burden for controlling pollution of water as percentage of turnover of the respective industries is given in Table 4.9 during the 1980s (based on respective Comprehensive Industry Documents).

Table 4.9
Annualised Costs of Pollution Control Relative to Turnover of Selected Industries

Small-scale paper industry using agricultural residue	below 5 per cent
Small-scale paper industry using purchased pulp and waste paper	0.55 to 1.19 per cent
Sugar industry	less than 1 per cent
Distillery (for two given schemes)	22.77 and 10.09 per cent respectively
Maltry	0.15 per cent
Brewery	0.91 per cent
Man-made fibres	1 to 2 per cent
Oil refineries	0.05 per cent at minimum turnover 0.01 per cent at maximum turnover

Source: CPCB, respective Comprehensive Industry Documents.

It would seem from the table that except for the distillery industry, the annualised cost burden is a small proportion up to 5 per cent, but mostly within 2 per cent. Even if the costs are passed on to consumers, it would not result in a great hike in prices and can be easily absorbed by the industry and consumers. It is only in the case of distilleries that the proportion is significantly high, being 10 to 23 per cent. The produce of distilleries is not of basic importance and its consumption is not something to be encouraged by passing on the externalities to the society. It is only reasonable that the industry and the consumers bear the cost of its pollution control, even if it is of a high order.

An earlier study of the costs of pollution control in Karnataka for the years 1982–83 to 1984–85 showed that the annualised costs of pollution control as a percentage of annual turnover of industries ranged from 0.02 to 7.4 per cent for all observations, and from 0.02 to 1.86 per cent for the modal quartile containing the bulk of observations (Ravichandran 1989, Chapter 6).

The study also showed that these costs did not affect the profitability of firms which was independently determined by other factors. Losses were made by some firms even where pollution control costs were low or declined. We quote from the study:

> Excluding the cases where losses were made, we have 24 observations pooled from 3 years, where the pollution control costs range from a mere 0.2 per cent of profits [Aeronautics] to 82.2 per cent [paper mill]. Dividing this range in four equal parts . . ., the bulk of observations viz. 20 out of 24 (83.3 per cent) are in the first quartile ranging from 0.2 to 20.7 per cent, with a median value of only 3.7 per cent. For the majority of the cases, pollution control costs cut profits only marginally (Ravichandran 1989, p. 164).

When pollution control costs are only a small proportion of the total costs and sales, they are not expected to affect profitability significantly. Our study of cement, textile and fertiliser industries in Karnataka for the year 1991–92 also confirmed earlier results and showed that pollution control costs are marginal. Under the case studies (in Chapter 6) we shall go into further detail, but at this stage, we note in Table 4.10 the costs per industrial unit on pollution control in relation to total costs and sales in the case of three cement factories, six textile factories and one fertiliser factory. It can be seen from the table that cost of compliance measured relatively in terms of percentage of annualised costs of pollution control to total sales was as low as 1.7 in the case of cement, 0.2 in the case of textiles and 1.9 in the case of fertilisers in Karnataka. If the cost of enforcement incurred by the SPCB is added to the annualised costs of pollution control and recovered from the respective firms, it makes hardly any difference. Cement and fertiliser are under 'Red' category while textile industry is under 'Orange'. While pollution control costs per cement factory on annualised basis was Rs 11.20 million, the inclusion of enforcement costs of Rs 16,359 would make it Rs 11.216 million. For fertiliser, it would be raised from Rs 31.54 million to Rs 31.556 million. In the case of the textile industry, it would be raised from Rs 0.53 million to Rs 0.5355 million.

Thus, one may conclude that even in a developing country, the costs of enforcement and compliance of pollution control appear on the whole to be easily absorbable, and will not make much difference to total costs and profitability. We had mentioned some limitations of our estimates of costs (mainly in respect of cost of enforcement). Even if these costs are doubled to make some

Table 4.10
Cost of Compliance of Pollution Control per Factory in Cement, Textile, and Fertiliser Industries (average of 3 years 1989–90 to 1991–92)
(Rs in million)

		Cement	*Textile*	*Fertiliser*
1.	Investment on pollution control	35.32	1.78	75.10
2.	Total investment	1,331.23	137.7	1,761.90
	Per cent of (1) to (2)	2.7	1.3	4.3
3a.	Fixed investment (annualised) on PC	7.77	0.39	16.52
3b.	Operating and maintenance costs on PC	3.25	0.14	15.02
3c.	Total annualised costs on PC	11.02	0.53	31.54
4.	Total annualised costs	405.11	38.54	651.91
	Per cent of (3c) to (4)	2.7	1.4	4.8
5.	Total sales	666.23	234.50	1,698.78
	Per cent of (3c) to (5)	1.7	0.2	1.9

Note: Annualised costs include 10 per cent depreciation and 12 per cent interest on total investment, and also maintenance and operation costs.
Source: Respective industrial units through questionnaires (three cement factories, six textile mills and one fertiliser plant in Karnataka).

allowance for a possible underestimation, they do not materially affect our conclusion.

4.5 PROSPECTS

The prospects of implementation of anti-pollution measures are not quite easy to see clearly due to conflicting influences in operation. On the one hand, as we have noted, the environmental laws have been made more effective during the 1980s, and the overall political mood in the country at higher levels is in favour of keeping the environment clean and healthy. The rise of popular movements in favour of the environment is a specially positive sign in this regard. The emergence of an indigenous industry to manufacture pollution control equipment is also a welcome trend. There are reported to be over hundred such manufactures in the country, and their design capabilities to address themselves to specific needs of industries are also improving. What is more, the cost of enforcement and compliance is within absorbable limits in most if not all industries. There are positive signs indicating hopeful prospects.

On the other hand, there are two developments which could be disturbing: (*a*) a relatively higher growth of more polluting industries vis-à-vis less polluting industries; and (*b*) economic reforms creating a mood for relaxation of regulation and control.

4.5.1 Growth of polluting vis-à-vis less polluting industries

The Government of India has identified 18 categories of industries as highly polluting. They include sugar, fertiliser, cement, distillery, aluminum, petrochemical, thermal power, caustic soda, oil-refinery, sulphuric acid, cluster of tanneries, copper smelter, zinc smelter, iron and steel, small-scale pulp and paper, dye and dye intermediate, pesticide manufacturing and formulation, pharmaceutical manufacturing and formulation.

The percentage rate of growth of these industries vis-à-vis others is presented in Tables 4.11 and 4.12.

We can see from Table 4.11 that the highly polluting industries are not predominant at present in terms of the number of industrial units, employment, value of inputs and outputs in Karnataka and the country as a whole. But they are predominant as fuel consumers in both Karnataka and India, and also as consumers of capital invested in the country as a whole. Fuel/output ratios and capital/output ratios are much higher for the highly polluting industries than for others. Higher pollution also goes with higher fuel and capital consumption. What is worrying is that their share in total industries is growing both in Karnataka and India. Table 4.12 shows that they are growing at a higher rate than other industries in all four respects, especially in terms of the number of units. This is disturbing because it makes the task of environmental regulation more difficult. It also reflects on the nature of growth taking place in India. An earlier study for the 1970s also showed that during that period the polluting industries had a higher rate of growth in terms of the number of units, value of input and value of output (Ravichandran 1989, p. 118). If the past trend continues during the 1990s, the prospects are not bright for the environment.

This is particularly so because even as it is, Karnataka's experience suggests that even in the case of these highly polluting industries, implementation of pollution control is not complete. As in April

Table 4.11

Percentage Share of Highly Polluting and Other Industries in All Industries for Important Indicators: 1980–81 and 1988–89—Karnataka and India

Indicators	*Karnataka*				*India*			
	Highly polluting industries		*Other industries*		*Highly polluting industries*		*Other industries*	
	1980–81	*1988–89*	*1980–81*	*1988–89*	*1980–81*	*1988–89*	*1980–81*	*1988–89*
Per cent share								
No. of industrial units	5.7	7.3	94.3	92.7	8.3	10.1	91.7	89.9
Total employment	31.5	27.3	68.5	72.7	27.1	26.2	72.9	73.8
Total capital invested	42.8	44.8	57.2	55.2	61.4	60.0	38.6	40.0
Fuel consumed	55.3	62.3	44.7	37.7	60.6	63.7	39.4	36.4
Value of input	29.8	34.6	70.2	65.4	35.3	39.6	64.7	60.4
Value of output	30.9	34.3	69.1	65.7	35.5	40.6	64.5	59.4
Ratios								
Fuel/output ratio (value of fuel per Rs 100 of output)	10.37	14.21	3.72	4.49	11.81	12.08	4.23	4.70
Fixed capital/output ratio	1.07	0.73	0.24	0.33	1.01	0.83	0.20	0.25

Table 4.12
Growth of Highly Polluting Industries vis-à-vis Other Industries (percentage change between 1980–81 and 1988–89)

Variables	*Highly polluting industries*		*Other industries*	
	India	*Karnataka*	*India*	*Karnataka*
i. No. of industrial units	31.6	34.4	5.7	3.2
ii. Fuel consumption	252.2	367.4	209.0	250.5
iii. Value of input	238.9	271.6	181.6	198.8
iv. Value of output	244.4	241.4	178.2	193.7

Note: The rate of growth in (ii), (iii) and (iv) is in money terms, not adjusted for inflation.
Source: Annual Survey of Industries for respective years.

1993, there were 99 units in the state under 18 industrial types identified as highly polluting. Of these, only 38 units are complying with the pollution control standards (source: KSPCB). The industries are such that in all of them, pollution control should have been implemented and operating.

4.5.2 Implications of economic reforms

Economic reforms introduced in India since July 1991 have been basically aimed at curbing inflation and balance of payments deficits, and promoting growth. These reforms included a package of measures to achieve these goals: cutting government spending and reducing fiscal deficits in relation to gross national product (GNP), encouraging exports through a variety of measures including devaluation, attracting foreign investments by dropping barriers to foreign entry, liberalising industrial production and privatisation of industries.

A concern has already been voiced in several quarters about the possible impact of these measures on poverty and environment (Kothari & Kothari 1992; Mukherjee 1993); we are concerned here with the environment, particularly with industrial pollution.

Fears have been expressed that in opening up the economy to foreign investment, there could occur a migration of dirty industries from the developed countries. This could be facilitated in a policy context where growth rates of GNP and foreign exchange earnings receive higher priority than implementation of environmental regulation. An additional complication is the assertion of the rights of the states in a federal set-up to have their own say in economic and environmental matters. The states, left to themselves, tend to take a more complacent look at environmental matters. An indication of this came from S. Bangarappa, who was the Chief Minister of Karnataka when the economic reforms were initiated in July 1991. According to a press report, he observed on the occasion of announcing a new industrial policy of the state in August 1991, that 'Clearances under Environmental Protection Act, Factories and Boilers Act and Labour Act, among others, would be suitably liberalised and streamlined to achieve the philosophy underlying the Industrial Policy Resolution' (*Deccan Herald*, 7 August 1991).

Fortunately, pollution control laws are not relaxed in the case of large- and medium-scale units, notwithstanding announcements like those of Bangarappa. But they are relaxed in the case of small-scale units. M. Arunachalam, the Minister of State for Small Scale Industry, Government of India, declared at a conference that 'SSIs should be left out of the purview of the proposed notification under Environment Protection Act 1986'. He felt that extensive environment controls should be applied only to projects above a certain size as long as SSIs did not undertake hazardous processes (*Deccan Herald*, 8 June 1993).

The SPCBs generally try to promote common treatment plants for SSIs and recover part of the cost from them. Where the SSIs are not located together, such facilities are not available and they are allowed to release the effluent into municipal drains or streams without treatment. Since the SSIs are growing rapidly, pollution caused by them is going to be a major problem.

According to press reports, the central government has now decided to delegate its powers to the states in granting environmental clearance to projects in certain critically polluting industries (*The Financial Express*, February 1993). This was due to all-round pressures from the states for greater decentralisation. In the wake of liberalisation of the economy, such decentralisation was an inevitable step but could be having undesirable consequences for the environment if the state governments start acting in a less than responsible manner. Though the Environment Protection Act (EPA) has given more powers to the central government in 1986, the post-1991 developments have complicated the matter to the discomfiture of the MEF at the centre.

A further worry is that under the urgency of cutting government spending to reduce fiscal deficits, the pollution control agencies—CPCB and SPCBs—may continue to receive much less funding than what they really need for effective working. This underlines the need for these Boards to raise their own finances through effluent and emission charges on polluting industries. The advantage of using proceeds of charges as the major source of finance for PCBs is that they provide an incentive for increased enforcement of antipollution law. In a policy context where market incentives receive more emphasis, there are brighter prospects of economic instruments of pollution control being made use of. The Policy Statement (Government of India, June 1992) has already mentioned the desirability of introducing fees or charges based on the discharge of pollutants. It is not clear if these will be based on total pollutants discharged irrespective of standards, or if pollutants within standards specified would be exempt. If there has to be an incentive for continuously improving technology and reducing pollution, the charge will have to be on total mass of pollutants. Since the working of economic instruments also needs close monitoring, the role of the Boards will hardly decline in significance even when the economic instruments are introduced. The prospects of regulatory instruments replaced by economic incentives are dim. On the other hand, it would be more realistic to expect them to work in a complementary way. They are expected to make the regulatory role of the Boards more automatic and easy, and reduce the power of discretion which the Boards still enjoy in administrating the laws.

The introduction of economic instruments will, however, require a simplification and relaxation of bureaucratic procedures. Bureaucratic hurdles can come in the way of a proper and effective operation of economic incentives. An example is tax relief provided to industries which have in-house facilities for generation of power. The industries have, however, complained that even when they are generating their own power, the tax relief is not given easily as a matter of course and bureaucratic delays have increased their cost.

As environmental awareness and responsibility increases in the public mind as well as industry, as has indeed been taking place, the future prospects are more optimistic. This is so in spite of industry-friendly ministers, politicians and technocrats. In an open democracy as in India, it is difficult to hide pollution and prevent political pressures building up against it. That indeed is the greatest hope for the environment.

Notes

1. In writing the country study on India including case studies, the authors have benefited from discussions with Dr K.C. Rathore of Central Pollution Control Board; and with Prof. B. Shivalingaiah, Member Secretary, and Mr Simha and Mr Sirisi, Board Engineers, KSPCB.
2. An Environmental Impact Assessment Report (EIAR) before giving NOC has been made mandatory for only 29 fragile projects through an Act 'The Environment Impact Assessment Notification dated 27 January 1994' and later Amendment Act on 4 May 1994 by the Ministry of Environment and Forests, Government of India.
3. As stated by Mr H.R. Sundareshan, Chief Manager, Commercial and Industrial Banking, State Bank of Mysore, Bangalore; and by Mr V.S. Reddy, Divisional Manager, Canara Bank, Bangalore.
4. During 1992–93, the proportion of 'Red' units covered declined to 29.8 per cent from 30 per cent while that of 'Orange' increased to 22.2 per cent from 22 per cent.

5

POLICY INSTRUMENTS FOR POLLUTION CONTROL IN THE NETHERLANDS

5.1 HISTORICAL OVERVIEW AND BACKGROUND

5.1.1 Structural backgrounds of Dutch environmental problems

With 393 inhabitants per sq. km (1987) the Netherlands has the highest population density of all OECD countries. Combined with a high per capita income, this results in the highest rate of all OECD countries for several pollution indicators per sq. km, while pollution per capita is around the average (RIVM 1991). The population is expected to increase from 15 million in 1990 to almost 17 million by 2015 (RIVM 1993).

By tradition the Netherlands has been an agricultural and trading nation. Industrialisation had a late start and only got under way in a comprehensive manner after 1945. The post-war period was characterised by a rapid expansion of the industrial sector combined with a striking increase in the share of trade in total output. Nearly all manufacturing sub-sectors contributed, but the chemical, petroleum refining and food processing industries showed the sharpest increases. This was caused not only by the discovery of natural gas but also by the growth of Rotterdam as an entrepot centre for oil, which led to downstream activities like refining and chemical industry. Much of the industrial capacity is concentrated in densely populated areas.

Export and import shares of GNP are large, 47 and 45 per cent respectively in 1990. Agriculture is an important export sector, with an export share of 25 per cent in 1990, but provides only 4 per cent of employment and 5 per cent of GNP (OECD 1993a). Agricultural production is tilted towards pollution-intensive activities, using large amounts of fertilisers and pesticides as well as (imported) feedstuffs. The most pressing problems in agriculture are the over-supply of manure, dehydration of soils, and contamination of soils and ground water by harmful substances. Energy intensity is relatively high, especially in horticulture, using huge amounts of natural gas for heating greenhouses.

The service sector's share in employment increased from 59 per cent in 1975 to 69 per cent in 1990 while the industrial share fell from 35 to 26 per cent. Transportation is an important service sector, reflecting the special geographic position of the Netherlands as 'gateway to Europe'. The fleet of trucks is large, encouraged by low prices for diesel oil, a strong pollutant when used by trucks (OECD 1990).

The government's current financial and economic policy is dominated by two objectives:

- To reduce the government deficit and the burden of collective spending;
- to reduce the unemployment rate, which is high by international comparison.

Presently, unemployment is a higher rating political issue than the environment.

5.1.2 Emergence of environmental awareness and policy

Concern about the natural environment has a long history in the Netherlands. The early initiatives taken by citizens and government were mainly directed to the directly perceptible kinds of damage and nuisance. The turn of the twentieth century witnessed the birth of several private nature protection associations. The Nuisance Act was enacted in 1875 in the days of the industrial revolution. The Nuisance Act was meant to protect man against danger, damage or nuisance from surrounding industrial activities. In 1896, the protection of the general state of health and safety was added to this Act. However, it was only after the Second World War that environmental problems became widely recognised as a major threat to man and nature.

By the early 1960s, the post-war reconstruction was complete. People were reasonably prosperous and more mobile than ever; they were well-insured for old age, against illness and infirmity, and were optimistic about the future. It seemed self-evident that there would be a further increase in prosperity, based on free enterprise and private industry, accompanied with a powerful government supporting economic growth and favoured by a stable parliamentary democracy.

In the second half of the 1960s, however, a wave of international cultural criticism developed. The rapid growth of population, expansion of consumption and production, urbanisation, greater mobility, increase of recreation and the shift in agricultural business led to a multitude of changes in the whole environment. At the same time, various signs of environmental decay appeared that led to a better recognition and analysis of all sorts of gradual environmental degradation. Water and air pollution, noise nuisance and an excess of waste soon became an acute problem (VoMil 1972). The publication of the report for the Club of Rome 'The limits to growth' (1972) drew attention to the finiteness of natural resources. The awareness of negative consequences of uncontrolled economic growth, including environmental degradation, led to the search for a different kind of prosperity.

In the 1970s, about 700 environmental organisations emerged. An important function of these NGOs was to focus on and deal with specific, often local, environmental problems. In contrast with the earlier established organisations, these groups were more politically-oriented, representing a wide spectrum of environmental and political ideas. These organisations found a broad support in society and can be characterised by a critical attitude concerning economic growth and technological development. In the 1980s, the government's attitude towards environmental movements changed, resulting in more cooperation between them. The policy of the environmental movement changed from a political to a more practical approach, searching for feasible solutions.

In the 1980s, groups like trade unions and industry began to realise the severeness of the environmental problems. Although many firms persisted in a defensive approach, the recognition that a change was needed and the willingness to change increased. This more cooperative approach contributed to the conclusion of more or less voluntary agreements between the government and trade and industry, often as a result of public pressure (see Section 5.2.2). The growing environmental consciousness, however, has not yet resulted in a fundamental change of production structures and consumption patterns.

5.1.3 Initial legislation and policies

The 1972 Emergency Memorandum on Environmental Hygiene is the first document in which the Dutch government presented a policy response to the environmental problem. Although it emphasised the connection between environmental process, a compartment-oriented approach was applied, aimed at the special features of water, air and soil. The Emergency Memorandum made an inventory of the most important problems per compartment and suggested an approach which leaned heavily on the permit system. In the 1970s, this approach resulted in detailed compartmental legislation. The Surface Water Pollution Act (1970) and the Air Pollution Act (1970) constitute the first generation of environmental Acts, completed by the Chemical Waste Act (1976), the Waste Act (1976), the Noise Nuisance Act (1979) and other Acts. The emphasis was on local, readily perceptible problems and public health.

The main regulating instrument used in the years after the Emergency Memorandum is the permit system, combined with a monitoring system. High expectations existed with regard to the enforcement of legislation. By introducing air pollution permits it was expected that '[i]n this way the industrial air pollution can be controlled as soon as the Air Pollution Act comes into operation' (VoMil 1972). A complicated permit system evolved, dispersed over a great number of authorities, which caused a great administrative burden for both firms and the public administration. The PPP, which is a leading principle till today, had been adopted in the Emergency Memorandum.

In the period after the Emergency Memorandum, environmental policy was confronted with a continuing appearance of new and more serious environmental problems, and ineffectiveness of the compartment-oriented permit system. The disappointing results were ascribed to three main problems:

1. The compartmental approach provided incentives to shift pollution to unregulated areas: e.g., air and water pollution was transformed into (chemical) waste.
2. The permit system was ineffective. For the authorities the permit system was very labour intensive (resulting in large arrears) and difficult to enforce, while effective sanctions were missing.
3. It turned out to be impossible to regulate emissions of all innumerable (potentially) harmful substances.

These failures stimulated the search for new policy principles.

5.1.4 Developments of approaches and principles

The disappointing results, arising from the compartmental and fragmentary approach, led to the development of two main principles of present-day Dutch environmental policy:

- dual approach;
- integrative approach.

The following two sections trace the significance of these two principles that contributed to the evolution of Dutch environmental policy.

Dual approach

The dual approach contains both effect-oriented and source-oriented policies.

The *effect-oriented policy* is directed at preventing and abating negative effects on environmental quality. Possible negative effects and the risk that these effects will occur are used to determine the maximum acceptable level of pollution, as well as the 'target value', at which no adverse effects occur at all. The effect-oriented approach cannot always be used to determine emission standards for individual sources. It is often impossible to 'translate' the environmental quality objectively into specific emission reductions because they are separated in space and time, and linked by intricate processes. Furthermore, even when this 'translation' is possible, the technology to achieve the required emission reductions may be lacking.

The *source-oriented approach* aims at abating environmental degradation at its source. It is effectuated by standards and permits for products and processes, generally based on available technology, and by taking into account their economic feasibility. A distinction is made between 'end-of-pipe' technologies, and 'structural source-oriented measures' to prevent the emergence of pollution, e.g., by a better design of the production process. With structural source-oriented measures, the production or consumption process is altered technically, so that emissions and waste flows are prevented without the use of add-on technology. These measures are also referred to as process-integrated environmental technology.

These two approaches supplement each other in Dutch environmental policy. Effect-oriented (environmental quality) *standards* determine the direction for long-term environmental policy while source-oriented standards are directly relevant for the individual polluter. Source-oriented preventive *measures* are generally preferred to (curative) effect-oriented ones, but are not suited to repair environmental heritages from the past (e.g., contaminated soils). Effect-oriented measures are also used to conserve valuable nature areas in situations of calamities, when they are much less expensive or when source-oriented measures take too much time.

Integrative approach

Integration of environmental policy relates to the coordination of environmental legislation and to the external and internal harmonisation of environment policy. The development of an integrated environmental policy is a process that started in the late 1970s and is still under way. In 1989, it resulted in an integrated policy plan (National Environmental Policy Plan, NEPP) and integrated legislation (Environment Protection Act, 1993). This section reviews the legislative, external and internal integration.

Legislative integration

When in the 1970s the government was confronted with the need for environmental policy, a legislative framework was missing. For practical reasons a quick compartmental building up of environmental legislation was chosen. The shortcomings of this system have been acknowledged at an early stage: legislation and procedures were too divergent, too detailed and not coherent. Implementation and enforcement of legislation failed.

The first legislative improvement was the enactment in 1980 of the General Environmental Provisions Act (Wabm). This Act integrated procedural rules of 13 environment Acts, and coordinated requests for permits. In the 1980s, other environmental laws have also been incorporated in the Wabm.

Because the environmental legislation was only partially integrated in the Wabm, in 1990 the government decided to integrate all legislations into one new Act, the Environmental Protection Act (Wet milieubeheer), enacted in 1993. Apart from integrating procedures, this Act also integrates in principle all environmental legislations step by step. Consequently the Act should provide the legislative basis for all measures needed for environmental protection.

Every four years an integrated National Environmental Policy Plan (NEPP), aiming at sustainable development, has to be drawn up. The NEPPs should contain a coherent strategy with respect to the various compartments distinguished, including the targets for the next eight years. A yearly programme has to evaluate the progress of execution of the NEPP and should adopt supplementary measures, if necessary. Objectives can be formulated in effect-oriented environmental quality standards which have to be observed by all governmental authorities.

An important result of the Environmental Protection Act is an integrated environmental permit which replaces five earlier compartmental permits. In addition, general rules and procedures of environmental policy as well as legislation are integrated, that provide the basis for other policy instruments.

External integration

In 1983, the government adopted a Policy Plan (VROM 1983) which made all other departments responsible for environmental concern as far as their policy has impact on the environment. As a result of this plan, coordination mechanisms are instituted in all departments to integrate environmental policy in their own policies. Integration is needed particularly for water management, conservation of nature and landscape and physical planning, for which several departments are responsible. The central government, provinces and municipalities have their own responsibilities, in particular, for town and country planning and water management. An institutional consultation structure has been created between the central government and (organisations of) these lower authorities.

Internal integration

Since the introduction of the Integrated Environmental Programmes in 1982, steps have been taken to organise environmental policy along several themes:

- Acidification: acidifying emissions that threaten aquatic and terrestrial ecosystems (first manifest in forests and lakes), and cause damage to materials and human health.
- Eutrophication: excessive supply of nutrients (phosphates and nitrogen) in surface waters and ground water, threatening water ecosystems.
- Diffusion: the spread of dangerous and harmful substances in the environment.
- Waste disposal: prevention, recycling and management of waste substances.
- Disturbance: dealing with (potential) disturbance by noise, stench and risks.

As environmental knowledge grew, these themes did no longer cover all environmental problems. In particular, global problems were added to the agenda. Therefore, the first NEPP (1989) introduced three new themes:

- Change of climate, including depletion of the ozone layer.
- Dehydration, concerning the results of ground water level decline.
- Squandering, addressing the use of exhaustible natural resources.

These themes have gradually replaced the former compartment-oriented approach. The themes are not only dealt with from a human health point of view, but due attention is given to the effects on ecosystems as well.

Policy principles

The following principles and a priori considerations have become cornerstones of Dutch environmental policy:

- the stand-still principle: prevention of further deterioration of environmental quality in areas that are still relatively clean;
- the principle of abatement at the source, when feasible;
- the PPP;
- prevention of unnecessary pollution;
- application of best (practicable) technologies: reducing the most polluting substances with best technical means (regardless of their costs), and less harmful substances with best practicable means (i.e., taking into account 'reasonable' costs, also known as the 'BAT (neec)' or 'alara' principle);
- isolation, management and control of non-processable waste;
- the dual policy: effect- and source-oriented.

In 1985, the idea of *social internalisation* was introduced as a new approach (VROM 1985a). Individuals and social groups should be aware of the environmental results of their behaviour and of their own environmental responsibilities. The disappointing experiences with enforcement of environmental legislation in the 1980s, combined with the tendency to simplify procedures, led to an emphasis on deregulation and self-regulation by the private sector, particularly by branch organisations and interest groups.

National Environmental Policy Plan (NEPP)

The first National Environmental Policy Plan (NEPP), published in 1989, is the policy response to the integral environmental survey 'Concern for Tomorrow' (Langeweg 1988). This survey was published by the National Institute of Public Health and Environmental Protection (RIVM) in 1988. It indicated that the emissions of several pollutants should be reduced by 70 to 90 per cent in order to achieve a sustainable environmental quality. It was published after the Brundtland report 'Our Common Future' (World Commission on Environment and Development 1987) which stimulated a further reflection on the way society exploits natural and environmental resources. These publications not only had a great impact on environmental policy but also contributed to a greater public awareness and the understanding that a thorough approach to the environmental issue is of national and international interest. It was acknowledged that intensification and extension of environmental policy was urgently needed. On the other hand, it was pointed out that it would take several decades to attain an environmental quality complying with the principles of sustainable development.

The NEPP is partly built upon the set of ideas embodied in the policy documents of the previous years. The recommendations of the Brundtland Commission add the following elements to those ideas:

- attention to the long-term effects for future generations;
- tackling of large-scale global environmental problems;
- mutual dependence of environmental quality and socio-economic development.

In the NEPP, the environmental effects of human activities are considered at various levels of scale: local, regional, fluvial, continental and global. Effects increasingly arise on higher levels of scale while sources are spread around the world. The perception of the relationship between cause and effect is weakened by increases in distance between sources and effects in time and location. 'Transfer' mechanisms shift environmental problems to other areas, scale levels or to the future, which is conflicting with the concept of sustainable development. Eventually, environmental effects will hamper human activities. A 'controlled feedback mechanism' has to be established instead of this 'natural feedback'. The effect-oriented feedback mechanism consists of preventing environmental damage by restricting harmful activities within an acceptable level of risk (feedback through risk management). The source-oriented feedback mechanism is directed at 'integrated life cycle management' (closing the substance cycles), energy saving, and quality improvement of processes and products.

The NEPP aims at achieving sustainable development within a single generation's time. In line with the Brundtland Commission, sustainable development is defined as 'a development that satisfies the needs of the present generation without endangering the possibilities of future generations for satisfying their needs' (VROM 1989). The NEPP contains the policy actions needed during the period 1990 to 1994, to get the desired development under way. The presented policy yields concrete targets and points of action, together with time-limits within which they must be realised. The related cost calculations are also drawn up. The policy to be pursued is considered along the themes mentioned before. A 'target group' approach is used, translating national policy in objectives for different sectors: agriculture, traffic and transport, industry, electricity and gas companies, the building trade, consumers, environmental companies, research, educational institutions and environmental organisations. To achieve the objectives for the target groups, covenants play an important role. The NEPP is co-signed by four (now five) ministers, which should guarantee the cooperation of the other departments to ensure external integration.

In 1990, a new government published the NEPP-plus, a supplement to the NEPP (VROM 1990). It contains policy intensifications, in order to realise the stated long-term objectives of the NEPP. Policy on instruments is adjusted, proposing a more important role for economic instruments.

In December 1993, the second National Environmental Policy Plan (NEPP-2) was published as it had become difficult to implement the policy measures of the first NEPP. The emphasis in NEPP-2 shifted from goal-setting to giving due attention to the target groups and taking measures to strengthen existing policy. However, the third National Survey on the Environment, published in 1993 by RIVM, showed that much remains to be done to meet the objectives, or even to go in that direction. For this purpose, NEPP-2 elaborates on measures for the various target groups (e.g., agriculture, industry, traffic).

5.2 INSTITUTIONS AND INSTRUMENTS FOR INDUSTRIAL POLLUTION CONTROL

5.2.1 Institutional organisation of environmental policy

Central governmental institutions

In 1971, the General Directorate of Environmental Hygiene was instituted at the Ministry of Public Health. In 1982, it was renamed

General Directorate Environmental Management (Directoraat Generaal Milieubeheer, DGM), and since that year it is part of the Ministry of Housing, Physical Planning and Environment (VROM). DGM has an important role in developing and implementing environmental policy. The main tasks of DGM are:

- to prepare, formulate and execute environmental policy;
- to supervise and enforce regulations.

The initial compartmental approach can still be seen in the organisation of the DGM's departments, of which some are especially responsible for one traditional compartment such as the departments for air, drinking water, water and soil. However, they have been supplemented by several other departments that take a more integrative approach. The Environmental Inspection (Inspectie Milieuhygiëne) is responsible for the enforcement of the environmental legislation. The most important supporting organisation is the National Institute of Public Health and Environmental Protection (RIVM), functioning as a planning office and research centre.

In addition to the Ministry of Housing, Physical Planning and Environment, there are other ministries responsible for other aspects of environmental policy. For the water compartment, the Ministry of Traffic and Public Works has most responsibility. The State Water Board (Rijkswaterstaat) comes under this ministry and is responsible for the quality of state waters. Concerning the conservation of nature, the Ministry of Agriculture, Nature Management and Fisheries has prime responsibility.

Advisory bodies

The Dutch political system is often described as a 'consensus democracy'. All important policy decisions are usually preceded by intensive consultations with all parties that may be directly or indirectly affected. Hence the strong preference for voluntary agreements (see Section 5.2.2) and the existence of several advisory bodies that may influence environmental decisions.

The Council for Environmental Management[1] advises the government on strategic issues such as objectives and principles of environmental policy, instruments and organisational matters. Several societal groups (including environmental organisations) are represented in this Council. Additional councils exist which deal with research in nature and environment (RMNO), regional planning (RARO), energy (AER), etc.

The Socio-Economic Council (SER) is an important body that advises the government on social and economic problems (including socio-economic aspects of environmental policy). In the SER, the trade unions, employers' organisations and independent experts are represented.

The Scientific Council for Government Policy (Wetenschappelijke Raad voor het Regeringsbeleid, WRR) is yet another advisory board. Its advices are based on independent scientific research rather than on sectional interests.

Provincial and local governments

There are three types of administrative bodies at the sub-national level: provinces, municipalities and water boards. These public bodies have autonomous tasks as well as co-governmental tasks. Often rule-giving competence is delegated by the formal legislator to these lower authorities. The Netherlands is divided into 12 provinces which have a small number of tasks in comparison to the central government or the municipalities. The provinces draw up provincial plans for environmental policy and are involved in granting permits. The tasks of more than 600 municipalities are quite extensive. The municipal responsibilities involve many environmental concerns such as physical planning, housing, traffic, recreation, levying several taxes and granting permits. The water boards are corporate bodies stemming from the thirteenth century. They fulfil specific tasks regarding the management of water: administering dams, drainage, and the cleaning of water courses. Many of them are also responsible for the purification of waste water.

European influence

The Netherlands is one of the 15 Member States of the European Union (EU).[2] Recently, the competence of the EU in environmental matters has been enlarged, although controversy remains about the demarcation line with national responsibilities. Most EU environmental legislation takes shape in so-called Directives which have to be incorporated into national law by the Member States. EU environmental policy is largely determined by the objective of an undistorted internal market. Therefore, environmental standards for *products* are being harmonised to a large extent. Differences in emission standards for *industrial sources* are less of a threat to the internal market. These standards are often only partly harmonised, allowing individual Member States to apply more stringent ones. The Netherlands has used its discretion on several occasions.

5.2.2 Instruments of environmental policy

Direct regulation plays a major role since the very beginning of environmental policy in the Netherlands. When in the 1970s environmental policies developed that covered the whole spectrum of environmental problems, direct regulation by means of permits and bans was the unquestioned approach. Originally, instrument choice was based on the principle that serious environmental problems should be regulated by means of direct regulation. Economic instruments should only be applied for less serious environmental degradation (VoMil 1974). In the spirit of the 1980s, the prominent role of the government became less self-evident, and the possibilities of a fine-tuned regulation by the government met with increasing scepticism. Environmental policy through direct regulation turned out to be far less effective than expected (Arentsen et al. 1993).

The policy response to this disappointing result was twofold:

- existing policy was intensified and extended by means of supporting measures to carry out, enforce and control regulations;
- alternative instruments were developed, such as communicative instruments, and some initial steps were taken to extend the use of economic instruments.

This section deals with the role and performance of direct regulation, communicative instruments and economic instruments, discussing the concrete measures taken.

Direct regulation

The core of instruments of environmental policy in the Netherlands consists of the prohibition to perform all sorts of environmentally harmful activities without a permit. Particularly in the 1970s, regulations have been enacted for all environmental compartments in which such a prohibition figured as a cornerstone. Permits have been encapsulated in a general framework. Several laws provided general rules and requirements to be effectuated by lower-level authorities which are responsible for carrying out this legislation. A number of permit systems have been intregrated in the Environment Protection Act (1993).

The permits contain particular prescriptions for an individual firm. These prescriptions can concern the targets as well as the means to reach these targets. In contrast with means-prescriptions, target-prescriptions leave it to the firms to decide by what means they will comply with the requirements. Many permits also oblige the firm to perform specified measurements.

Apart from permits, other direct regulative instruments are being applied, such as bans on certain (applications of) substances and products, as well as general instructions and standards for processes and products. This type of regulations is largely based on the Environmentally Hazardous Substances Act. In order to avoid trade barriers within the EU, there is a tendency towards harmonisation of product- and substance-oriented measures at the EU level. Nevertheless, the Netherlands sometimes takes more stringent national measures.

Environmental Impact Assessment (EIA) can also be regarded as an instrument of direct regulation. The obligation to perform an EIA is a condition to obtain a permit for several kinds of large-scale activities and investments with considerable (potential) environmental impacts.

The permit system

The Nuisance Act aimed at restricting excessive danger, damage and nuisance caused by firms, through a permit system executed on the municipal level. However, the increasing environmental problems on a larger scale and pollution by mobile sources could not be dealt with effectively in this way. Therefore, supplementary legislation was necessary. This resulted in a series of laws, aimed at distinguished environmental compartments, such as the Surface Water Pollution Act (1970) and Air Pollution Act (1970). The compartmental approach involves the risk of replacing one kind of pollution by another. Integration and attuning the procedures and contents in decision-making was meant to prevent this effect.

The compartmental laws all had the character of framework acts: specifying the essential prescriptions was largely left to the (numerous) executive authorities. General indicative policy objectives were missing and binding objectives could be found even less. This was partly due to a lack of knowledge and the wish for flexibility. The permit system, carried out at a decentralised level and binding the permit-granting authority only slightly to higher prescriptions, gave the executing authorities the possibility for a more subtle influence on environmental behaviour, adjusted to the local situation. This, however, also created the possibility for the authorities to weigh interests other than environmental ones more than they should, under pressure of local circumstances. Requirements became to be negotiated and evaded. Decentralisation resulted in a shift of power to smaller and consequently less powerful authorities. The framework Acts gave them great autonomy but failed to give them support in the struggle against numerous other interests (Glasbergen 1989).

The application of the 'best practicable means' principle amplified this effect, leaving enough space for negotiations about what should be regarded as 'economically practicable'. Only for some exceptional risks, the alternative and more objective 'best technical means' principle was applied. In many cases disproportional costs have been compensated by state subsidies.

An important result of the Environment Protection Act is the replacement of five compartmental permits by one integrated permit. In this integrated permit, a broader concept of 'environment' is used: it includes waste disposal, energy and resource use, and related transportation. The Nuisance Act has been replaced by legislation of the Environment Protection Act.

Permits have to be based on the NEPPs, general rules, policy decisions and standards. In addition several directives can be used, mostly originating from the government, but also from quasi-governmental organisations.

The 'alara' principle is included in the Environment Protection Act but leaves space for interpretation (Brussaard et al. 1993). A firm can be obliged to investigate the possibilities for less detrimental production processes. The separate permit of the Surface Water Pollution Act is continued. Requests and issuings have to be combined with the integrated permit.

The Environment Protection Act also includes the possibility to issue general regulations. Three kinds of general regulations can be distinguished:

- General regulations that replace the obligation for a permit. For many firms (especially in relatively 'homogeneous' branches of trade and industry), it is regarded to be more efficient to apply standard regulations instead of permits. Out of the 400,000 institutions that potentially fall under the permit obligation, 300,000 are regulated by these general regulations. In this way the permits did become a less generally applied instrument (Brussaard et al. 1993);
- General regulations, supplementary to the permit are meant to regulate only certain aspects;
- Instructions directed to the lower, permit-issuing authorities to include certain prescriptions in the permit. The advantage of this instruction is that all obligations for a firm can still be regulated in one document: the integrated permit.

All procedures concerning permits are included in the Environment Protection Act. The procedure is very detailed and gives citizens the possibility to object to the permit. Generally the municipalities are responsible for deciding the permits, in a few cases the provinces or the Minister of the Environment is responsible. The decision about a permit should be taken within six months.

An evaluation of the performance of the Environment Protection Act cannot yet be made, but many shortcomings of former legislation seem to be repaired. Why this repair was needed will be illustrated in the next section that evaluates the practice of the former permit system which will be step by step replaced by the Environment Protection Act.

Evaluation of the permit system

A rough estimation, summarising several evaluation reports, states that with direct regulation about half has been achieved of what the government intended (Bressers 1991). Several evaluations have been made of the main permit systems that resulted from the Nuisance Act, the Air Pollution Act, the Surface Water Pollution Act, and several regulations concerning harmful substances or radiation.

The first study about the performance of permit systems concerned the Nuisance Act (Arentsen et al. 1993; Glasbergen 1989). The results of this evaluation came as a great shock. It was estimated that in 1976 only 25 per cent of the firms obliged to have a permit indeed had a valid permit. The procedure to obtain a permit generally took about one year. Because most of the firms were operating before a permit was required, issuing permits was not considered urgent and responsible departments were poorly equipped. In 1989, the percentage of firms with a permit had increased to 38 per cent out of about half a million firms that came under the Nuisance Act. Twenty-four per cent of the firms had adequate permits. Various firms that caused no or very little environmental damage were obliged to have a permit while several large polluters were not. Inspection of firms was often very infrequent.

Milieuvoorschriften (1981–93) ascribes the poor performance of the Nuisance Act to the following factors. The formal sanction, i.e., closing down the firm, is seldom used and appears to be only symbolic. Sources of nuisance often result from irreversible investments. Enforcement by inspection has no priority. Economic interests, in particular employment, often outweigh an effective implementation of the permits. Seemingly, social side effects (public consciousness) and informal consultations are the only reasons why the Nuisance Act still has any effect. In 1981, a penalty was introduced as an additional sanction but it is hardly used due to stringent juridical requirements. The actual situation is that the environmental circumstances in many firms still do not comply with the prevailing prescriptions (Arentsen et al. 1993).

Four years after the enactment of the Air Pollution Act, about 50 per cent of the firms covered by this Act had a permit. After eight years this percentage had increased to about 75. An evaluation in the industrial Rijnmond area found a reduction of SO_2 between 1974 and 1980. This could, however, not be ascribed to the permit system, because the permits allowed the SO_2 emitting firms (mainly refineries) to emit one-and-a-half times as much as they actually emitted (Arentsen et al. 1993). The reduction was caused by lower economic activity, use of cleaner fuels, and a decrease of the legally allowed sulphur grade of fuels (Twijnstra Gudde 1982). The provinces are primarily responsible for issuing air pollution permits. An evaluation in 1992 (Arentsen et al. 1993) showed that monitoring air quality standards was done far less than required, and in 50 per cent of the cases reporting did not take place.

The issuing of permits within the framework of the Surface Water Pollution Act also shows large arrears, and enforcement is often inadequate (Arentsen et al. 1993). The prescriptions which the water boards include in the permits are sometimes less stringent than intended by the national policy makers. For example, even for 'black list' substances like heavy metals for which the 'best technical means' should be used, they tend to take the economic situation of the firm into account ('best practicable means').

To conclude, it can be stated that the main instrument of direct regulation—the permit—is still not functioning very well despite improvements in recent years. Firms often do not have an adequate permit, or they do not comply with the conditions of the permit. Authorities often have to tolerate this, as they do not have the resources to provide all firms with up-to-date permits and to perform effective enforcement. Presently, non-compliance situations can be formally sanctioned on a temporary basis by means of a 'tolerating order'. The government's objective is, however, to ensure that all firms have adequate permits, and that they comply with them. 'Tolerating orders' should only be issued in exceptional cases.

The relatively large autonomy for the permit-issuing authorities may lead to less stringent demands than needed from a national environmental policy point of view. Under the new Environment Protection Act, the central authorities have more possibilities to influence this, i.e., by means of general regulations.

Several environmental problems, especially those caused by 'non-point' sources, cannot be dealt with by means of the permit system. In these cases, direct regulation would mean bans or limitations on production, trade, use or disposal of certain products, or on other activities. For dangerous substances such as PCBs, heavy metals, pesticides and CFCs, this is probably the only effective and acceptable instrument. In other cases, where a zero emission is not necessary or feasible, the use of other instruments should be considered.

Communicative instruments

Due to the lack of confidence in effective regulation, in the 1980s the method of governance of environmental problems showed a reorientation towards a more prominent role for individual responsibility, in particular, of trade and industry. The original concept of 'social internalisation' of environmental responsibility has been developed into a 'target group approach' and emphasis laid on information and education. By means of consultation and negotiation with organised parts of trade and industry, attempts have been made to utilise and stimulate the responsibility of the target groups themselves. For this approach communicative instruments have been applied more intensively. Three elements can be distinguished: education, covenants and the environmental management system within firms.

The government promotes *education* on nature and environment through its own activities and through subsidies to environmental organisations. In 1990, a publicity campaign was launched to stimulate environmentally responsible behaviour. Households are informed about possibilities to reduce consumption of energy and water. An environmental labelling system for relatively environment-friendly products has been introduced in 1993. According to the environmental advertising code, environmental claims in advertising have to be provable. Information supplied to and by firms sharply increased in the last few years. Firms increasingly need information about technological possibilities for energy saving, waste prevention and treatment, integrated environmental management and their internal environment management system. Several environmental centres were instituted, supported or initiated by the government, e.g., a network of innovation centres and environmental services for firms. The 'Prisma' project, initiated by the government, analysed possibilities to prevent waste and emissions. In many cases waste reduction proved to be cost-reducing (WRR 1992). Using this kind of instrument in the Netherlands is strongly facilitated by the obligatory registration of firms and organisations.

Initially, *covenants* (agreements between government and firms) were focused on substances or products. Examples are covenants on phosphate-free detergents and the banning of CFCs. The objectives of these agreements have generally been fulfilled, but it is unclear whether the results are really due to the covenant, as many other factors may have contributed (Klok 1989).

A new kind of covenants are agreements with firms or branches about their part in realising environmental policy targets. These more comprehensive covenants play an important role in the implementation of the environmental policy as formulated in the NEPPs, especially for the industrial sector.

The government distinguished 15 industrial branches and intended to negotiate agreements with each branch. These 15 branches together are responsible for 90 per cent of the industrial burden on the environment (KPMG/IVA 1993). This 'target group' approach is carried out in several steps: in cooperation with the branch organisations and public authorities, for each of the selected branches, the government aims to make a declaration of intent containing an 'integrated environmental task'. An important element is making the firms conscious about their own responsibility. The next step depends on whether the branch is homogeneous or heterogeneous. For homogeneous branches an implementation plan is made at branch level and elaborated in concrete actions. This plan, made in collaboration between the branch and public authorities, holds for the entire branch. For heterogeneous branches the integrated environmental task is elaborated at the level of individual firms. The permit-issuing authorities are involved in laying down the so-called environmental business plans. This package of measures is based on an assessment of achievable reductions, considering the firm's viability. An evaluation is made after four years which can result in a supplementary plan.

Recently, a number of disappointing results gave rise to criticism regarding the merits of covenants (Houtsma 1994). In 1991, a covenant was concluded concerning packaging. It was agreed that the volume of packaging materials would be reduced with 10 per cent in the year 2000 compared to 1986. In 1992, however, the volume of packaging used increased by 27 per cent, and only one-third of the firms are actually working on the execution of the covenant. The declaration of intent with the metal industry shows a similarly disappointing result. The emissions of acidifying and harmful substances are expected to be at the most 33 per cent less by 2000 than in 1985 instead of the 55 to 75 per cent that was agreed upon.

These results illustrate that combining flexibility for firms and effectiveness for the environment is a difficult job. Environmental organisations, the Ministry and lawyers oppose 'soft' agreements while companies plead for flexibility without detailed governmental regulation. To improve the performance of covenants it is suggested to make them an element of a broader package of instruments (Houtsma 1994). The advantages of covenants—faster policy making, greater acceptance, flexibility and deregulation—should be accompanied by the prospect of sanctions and direct regulation in case of non-compliance. Therefore, the legal status of covenants has to become clear. This does not, however, solve the lack of parliamentary democratic control nor does it guarantee the social acceptance of covenants.

The relationship between voluntary agreements and direct regulation in Dutch environmental policy is complex. The government wants to make it attractive for firms to accept voluntary commitments. But at the same time, it has little to offer in return. Generally, the targets and measures agreed upon will have to be laid down in some form of regulation in order to enforce compliance and to prevent free-riding. One way of stimulating the preparedness to participate in a voluntary agreement might be to impose more stringent regulations and/or more frequent checks on non-participants.

It is difficult to make an exact assessment of the effectiveness of covenants. Considerable results have been achieved, but in several cases this could be ascribed to other factors than the covenant itself (Klok 1989). In other cases, the results were less than what was agreed upon or was needed. It can be questioned, however, whether these goals would have been achieved by applying only regulatory instruments, and at what price (Arentsen et al. 1993).

Apart from 'target group' consultation and covenants, *environmental management systems within firms* are another way to contribute to the 'social internalisation' of environmental concern. In consultation with the industry in 1989, objectives have been formulated by the government. For the year 1995, the target was for 10,000 to 12,000 firms (large polluters, or firms causing special environmental risks) to have a complete environmental management system. Additionally, the government expected all relevant branches to take steps to introduce a partial system before 1995, to be implemented by about 250,000 firms (Boersema et al. 1991).

An environmental management system consists of the following elements: environmental goal setting and action programme, integration of environmental concern within the organisation, measurement and registration, education and training, and control and report. The government considers industry responsible for the introduction of environmental management within firms. As yet firms are not obliged to introduce an environmental management system. A subsidy is available for projects to gain experience with such a system.

Although the targets have not been attained, the introduction of environmental care systems is well under way. Major enterprises are more likely to live up to expectations than small ones (OECD 1995).

Economic instruments

The 1980s have seen a rising interest in economic instruments, as more cost efficient approaches were sought in the face of budgetary stringency. In addition, economic instruments have come to be seen as facilitating environmental innovation. These instruments have been debated and studied extensively but they play as yet a limited role in Dutch environmental policy. In the Netherlands, like in almost all countries, economic instruments are regarded only as secondary means to reach certain environmental goals. From the outset, direct regulations have been supplemented by economic instruments, mainly user and product charges, tax differentiation and subsidies. By international standards, economic instruments play a significant role, with the relative burden of product charges in the Netherlands being the highest in the OECD area (OECD 1990).

Figure 5.1 shows charge revenues and their composition in the period 1975–93. According to the NEPP-plus, in 1994 charges would finance 37 per cent of the costs of environmental policy. Another 49 per cent would be financed directly by the private sector while public expenditures would account for the remaining 14 per cent. In 1988, this distribution was 29 per cent for charges and 36 per cent for both direct private expenditures and public expenditures.

Figure 5.1: Composition of Charge Revenues

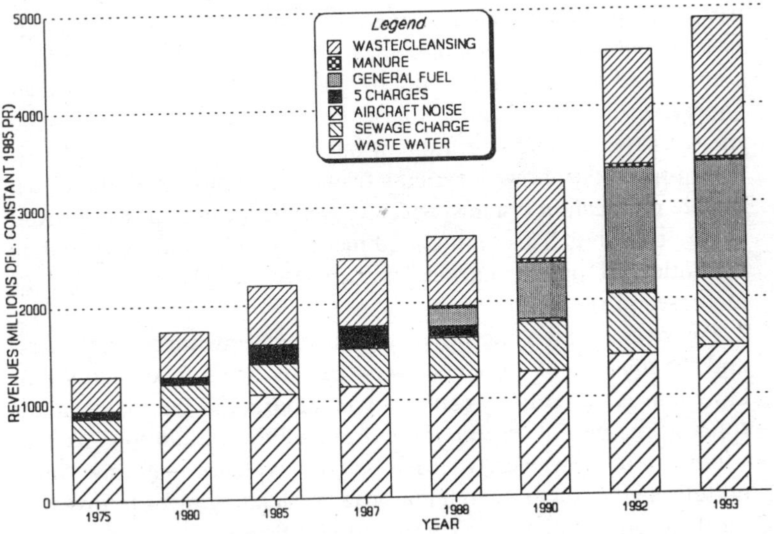

Charges can be applied for roughly two reasons:

1. The incentive function: to reach environmental goals when direct regulation or other instruments cannot be applied or obviously have great disadvantages;
2. revenue raising for specific or general (environmental) expenditures.

The environmental charges presently applied in the Netherlands have primarily a revenue raising function. The level of the charge rates is usually too small to provide an observable incentive effect, with the notable exception of the water pollution charge.

The following discussion deals with the most important charges which are being used or discussed in the Netherlands.

The *water pollution charge* (Surface Water Pollution Act) is by far the most important effluent charge in the Netherlands. It has a proportional rate on waste water disposal by large firms while households and small firms pay a fixed rate. No distinction is made between firms which comply with prescribed effluent limits and those which do not. Neither is a higher rate applicable to amounts of waste water which exceed the limits than for those which remain within them. Monitoring is restricted to large polluters. The charge base includes bio-degradable matter, suspended solids, toxic substances and heavy metals. Initially the system was set up as a user charge system. The Water Boards use the revenues for financing the construction and operation of public water purification plants. As it turned out, however, many firms decided to install their own private purification systems, which, given the levies imposed, proved to be profitable. This unintended incentive effect led to an under-utilisation of public purification capacity, which in turn led to increases in the charge rates.

Figure 5.2 shows the development of the average charge rate and the amount of effluents discharged by firms. The decrease in emissions is not entirely caused by the charge, as the figure might suggest: the permit system of the Surface Water Pollution Act also contributed. However, there is strong evidence that the water pollution charge has been a major factor leading firms to invest in purification installations (Bressers 1983; Schuurman 1988).

The visible seriousness of the water pollution problem, the transparency of allocation of revenues for purification and increasing charge rates are regarded as key factors in the high effectiveness of this charge (Opschoor & Vos 1989).

The charge level for heavy metals has been too low to provide an incentive to reduce emissions of heavy metals by industry (Bressers 1983). This is a result of the choice to base the charge level on purification costs and not on the desired level of pollution.

Effluent and emission charges are generally effective, generating strong incentives for emission reduction, provided that the level of the charge is high enough to make investment in abatement technology profitable, e.g., the waste water charge. This charge, however, also demonstrates that the effect of charges is hard to predict and might be unexpected or even undesired, which has to be taken into account when charges are applied as alternatives to direct regulation. For many types of pollution, e.g., air pollution, monitoring devices may be too costly to be implemented. In those cases, product charges may be more efficient.

The *general fuel tax*, introduced in 1988, is primarily a revenue raising charge, but can also have incentive effects as firms can

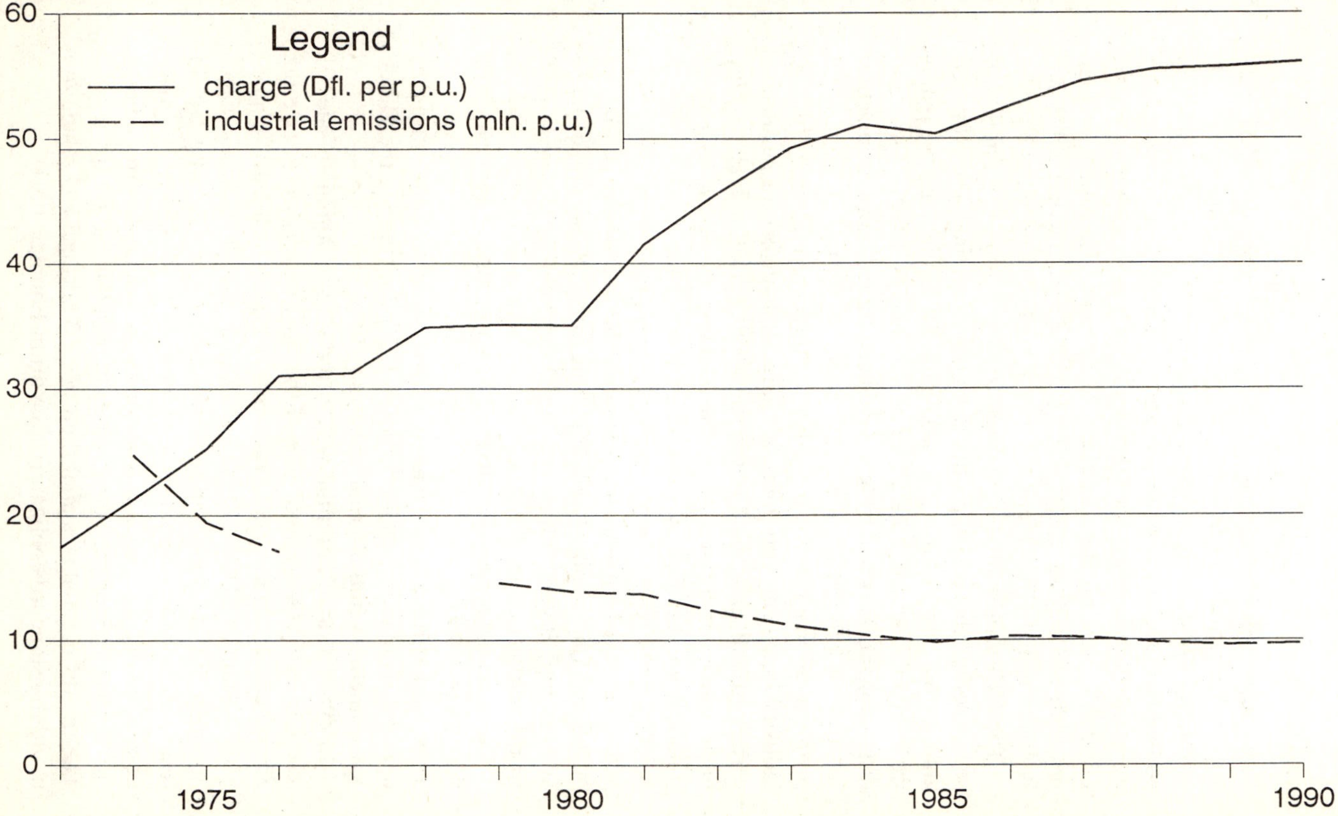

Figure 5.2: Development of the Average Charge Level and Waste Water Emissions by Firms

obtain a rebate when they apply certain abatement technologies. The charge is levied as a surtax on existing excise duties on fuels and is differentiated according to approximate pollution contents of different fuels. Originally the revenues of this tax were meant to compensate for environmental damage. The revenues were, however, increasingly used for other ends, such as for public transport or for the general state budget. At present, the formal linkage between the fuel tax and the budget for environmental policy has been abolished. The revenues are now part of the general budget.

The NEPP presented a proposal to investigate the feasibility of several other charges on products. In the mean time, the proposed charges on certain kinds of packaging, PVC, light bulbs, paints with organic solvents, metal degreasing agents, blasting grit, and disposable crockery and cutlery have been turned down for practical reasons. In some cases, alternative solutions for the environmental problem at hand were arrived at, after consultations with the industry involved. The plans of charges (or taxes) on pesticides, waste, ground water, and raw materials like sand, gravel and clay have until now survived. The introduction of an energy charge (apart from the already existing fuel tax) should, according to NEPP-2, preferably take place in a European framework. However, if the EU fails to reach an agreement, the energy charge may be introduced on a national scale, but only for small users (in order to avoid competitive disadvantages for Dutch industry exposed to the world market).

Two past examples of a successful environmental application of the tax instrument are the *tax differential* on car prices and the associated surcharge on leaded petrol. In both cases stringent national regulation was hindered by EU rules, forcing the Dutch government to find alternative ways.

In 1986, a tax differential was introduced to stimulate sales of 'cleaner' cars. The general rate of the existing special car sale tax was raised while a reduction was given for cars that already complied with announced EU-1993 standards. A major information campaign, in which the car trade and industry cooperated, played a crucial role to overcome consumers' hesitation. In 1986–92, the effectiveness of the tax differential was above the government's expectations: the tax rate was adapted downwards to maintain budget neutrality. The tax differential was abolished in 1993 because EU regulation made the standards obligatory (Savornin Lohman 1994).

To facilitate the introduction of catalytic converters in cars (that need unleaded petrol), and to reduce lead emissions from car traffic, in 1986 a tax differentiation was introduced between leaded and unleaded petrol. An 'industrial agreement' with regard to using the price difference to abolish leaded 'normal' petrol was reached between the authorities and oil companies. The measure had the effect of a ban. The substitution of one kind of petrol for another prevented extension costs for filling stations and petrol companies. The market share of unleaded 'super' petrol rose steadily from below 1 per cent in 1986 to about 78 per cent at the end of 1993,[3] while 'normal' petrol has been removed from the market.

The effectiveness of this instrument has been strongly supported by:

- already existing structures (excises on petrol);
- budget neutrality;
- agreement with the homogeneous and localisable group of oil companies and importers who carry out the measure;
- a direct relation to the problem and the behaviour of the target group;
- low costs of execution (Klok et al. 1992).

This combination of favourable circumstances is not likely to occur frequently.

Although *subsidies* are basically at odds with the PPP, they are still being applied on a fairly wide scale. The EU and the OECD allow certain exceptions under the PPP, under specific conditions.

Subsidies are presently used in Dutch environmental policy in order to facilitate the acceptance of stringent policy measures and to promote research and implementation of clean technologies and pollution control equipment. Financial compensation can be obtained when firms have to comply with direct regulation which results in substantially higher costs than those incurred by other comparable firms, due to exceptional circumstances. Furthermore, over the past few years an accelerated depreciation scheme is operational, under which investments in certain specified environmentally-friendly capital goods can be entirely deducted from taxable income in the year of purchase. For this scheme, Dfl. 120 million per year is available.

Theoretically, *tradable permits* combine the allocative efficiency of an emission charge with the certainty of an emission ceiling associated with direct regulation. In practice, tradable permits have only been applied on a considerable scale in the United States. In the Netherlands, this instrument has been discussed and proposed on several occasions but its acceptance and feasibility remain low. Nevertheless, the NEPP-2 announces new research into the possibilities for the use of tradable emission permits in Dutch environmental policy.

5.3 EFFECTIVENESS OF PAST AND PRESENT POLICIES

Since the NEPP (1989), the objectives of the Dutch environmental policy are related to eight policy themes. Each theme is elaborated into several indicators, for which quantitative aims have been formulated. Objectives and realisations for six themes are presented in Figure 5.3. It can be stated from Figure 5.3 that Dutch environmental policy has succeeded in breaking the linkage between GNP growth and increasing environmental pressure. At the same time, achieving the policy objectives for the year 2000 will require additional substantial efforts, not to mention the efforts needed to achieve the long-term sustainability targets.

This section will try to assess the effectiveness of Dutch environmental policy by means of brief descriptions of the policy themes, the policy objectives under these themes, and the results which have been achieved until now.

Climate change

The climate change theme includes the intensification of the greenhouse effect as well as damage to the ozone layer.

The greenhouse effect

The man-made part of the greenhouse effect is brought about by the emissions of carbon dioxide (CO_2, energy, deforestation), methane (CH_4, agriculture), dinitrogen oxide (N_2O, agriculture), chlorofluorocarbons (CFCs, industrial products) and ozone (O_3). Carbon monoxide (CO), nitrogen oxides (NO_x, energy), and

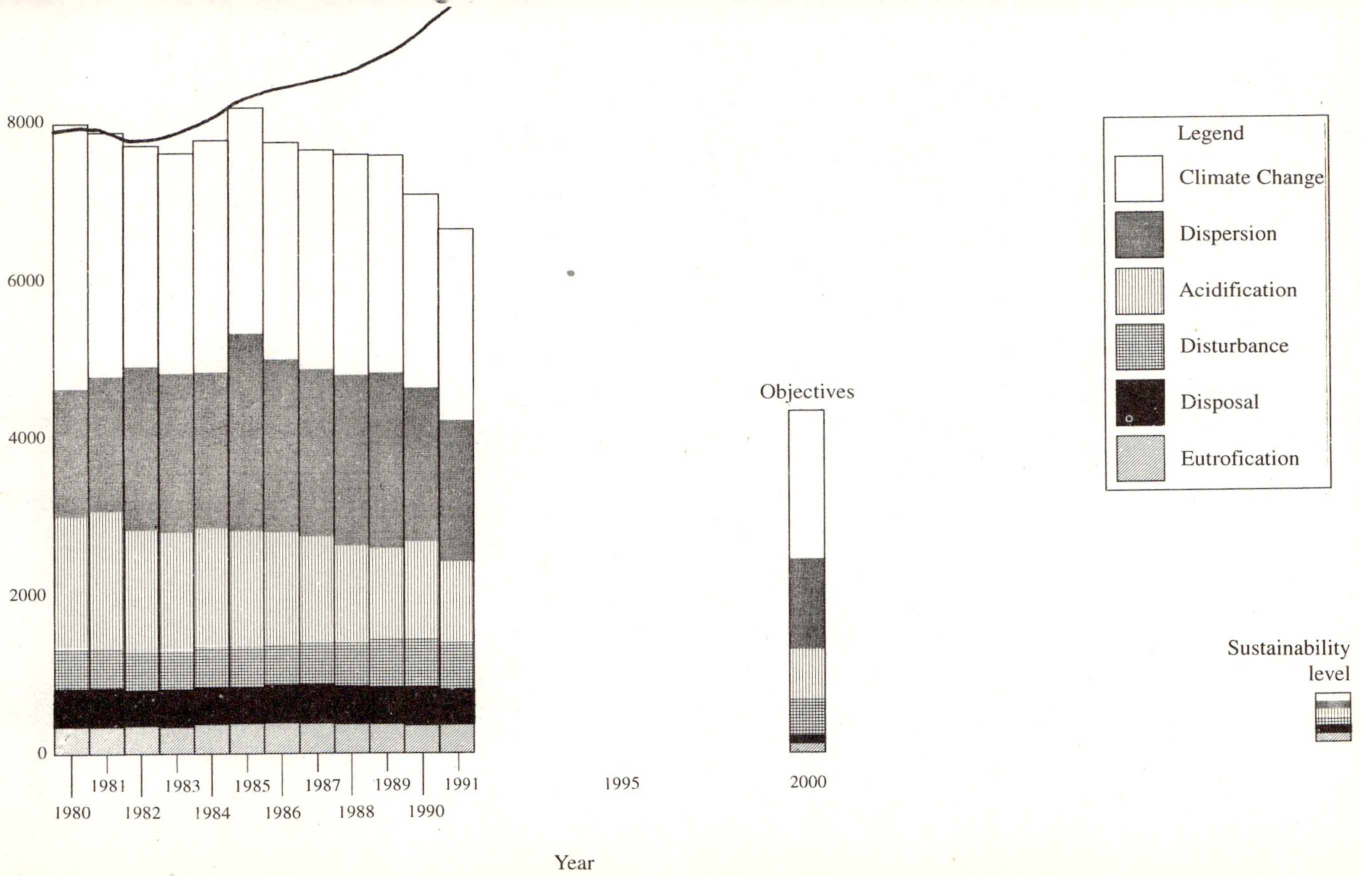

Figure 5.3: Environmental Indices, Targets and GNP Growth in the Netherlands

volatile organic compounds (VOCs) have an indirect effect (RIVM 1991). The Dutch contribution to global emissions is less than 1 per cent in absolute terms but above average in relative terms (for example per inhabitant). In 1988, industry contributed 39 per cent to the greenhouse gas emissions and 19 per cent originated from the energy sector (RIVM 1991). Objectives for CO_2, CH_4, CFCs and N_2O are formulated under the climate change theme. Emissions that primarily have an acidifying impact are dealt with under the acidification theme.

Since 1982, CO_2 emissions are increasing constantly by about 2 per cent yearly. CO_2 emissions increased by almost 11 per cent in the period 1989 to 1993 (CBS 1994c). The NEPP aimed at stabilisation by 1995 at the 1989 level, and a reduction by 3 per cent by the year 2000. Dutch CH_4 emissions in 1991 were up 12 per cent compared to 1980 (Adriaanse 1993). The increase is mainly attributable to agriculture. Other target groups that cause CH_4 emissions are the energy sector and the waste removal sector. It is assumed that N_2O emissions have stabilised during the last decade. The policy target for the years 1995 and 2000 is a further stabilisation at the 1990 level.

Between 1991 and 2000 the total emission of greenhouse gases in the Netherlands should decrease by 5 per cent (in CO_2 equivalents). Measures to realise this target include energy saving, 'better' fuel utilisation, traffic and transport measures and waste reduction. The use of coal will be curbed while the use of natural gas and imports of electricity will rise.

Energy production and consumption are responsible for about 50 per cent of the emissions of greenhouse gases. To achieve a 20 per cent increase in efficiency in energy production by 2000, the share of coal-fired power plants has to be diminished. In addition, energy efficiency should improve by 2 per cent per year. This is to be achieved by an extensive information campaign, coordination between governmental organisations and energy producers, minimum efficiency standards for equipment and buildings, research, subsidies and fiscal measures. For industry, the government aims at a 20 per cent increase in efficiency in energy use by 2000. Multi-year agreements have been concluded by the government, energy producers and industrial sectors. In addition, the government gives financial support to the development and introduction of new technologies. In consultation with the government, the energy distribution companies agreed upon an action programme on CO_2,

SO_2 and NO_x. This programme is financed by government subsidies and by passing on the costs to consumers. A subsidy to consumers for insulation and more efficient central heating is financed by the Ministry of Economic Affairs and the energy distribution companies. Experience shows that by means of high energy prices, an annual 2 to 3 per cent improvement in energy efficiency can be obtained. Combined with an economic growth of 2 to 3 per cent, CO_2 emissions would stabilise. The possibility of continuing efficiency improvements with the currently low energy prices is questioned (Janssen et al. 1991).

Proposals to introduce an incentive charge on fuels met strong opposition, especially from the agricultural sector and employers' organisations, because of the assumed risks for their international market position. The Dutch government would have preferred an EU-wide energy tax, in order to mitigate negative economic impacts. However, lacking agreement at the EU level, an energy tax was introduced in 1996, effectively affecting small users only.

Depletion of the ozone layer

The compounds most damaging to the ozone layer are CFCs and halons. The objectives of Dutch environmental policy are in line with the Protocol of Montreal, which stipulates worldwide eradication of the use of these substances by the year 2000.

In 1990, the government decided to strive for a sharp decrease in production and consumption of CFCs, to only 6 per cent of the 1990 level by 1995. By 1992, more than half of the reduction had been reached by means of a CFC action programme agreed upon by the government and industry in 1990. A complete phase out of CFC use had been achieved by 1995 (VROM 1996).

Halons were also phased out completely by 1995. The halons in use have to be recycled. The chief source of halons consists of fire-control agents. It is estimated that the production of halons in 1991 has decreased by 43 per cent compared with 1980 (Adriaanse 1993).

Acidification

The Netherlands emits relatively large amounts of acidifying pollutants. It is also among the countries in Europe that is suffering most from acidification. In the 1980s, the Netherlands' export of acidifying substances was about three times the import of these substances (VROM 1987).

The acidification theme includes sulphur dioxide (SO_2), nitrogen oxides (NO_x), ammonia (NH_3) and volatile organic compounds (VOCs). Emission of NH_3 is of great relevance for the Netherlands as it has the highest NH_3 emission per hectare in Europe (VROM 1985b).

Deposition standards, critical level

In 1986, the long-term acid deposition objective was determined at 1,400 acidification equivalents per hectare per year (Aeq, H+ equivalents; VROM 1985a). At this level of deposition, the most serious effects of acidification are prevented. In 1987, acid deposition in the Netherlands averaged 5,300 Aeq (VROM 1989). A policy evaluation in 1987 showed that objectives concerning SO_2, NO_x, NH_3 and VOCs would not be met by the year 2000 due to higher economic growth, an increase in the use of cars, and an expansion in the number of livestock. Furthermore, nature was found more vulnerable than assumed. In 1988, it turned out to be necessary to adjust the standard to 400–700 Aeq as the critical level (RIVM 1988). At deposition levels below 400 to 700 Aeq, only negligible effects occur.

Initial policy

As the Emergency Memorandum (VoMil 1972) was mainly concerned with human health, and acidification was unknown as such, measures were taken to abate and diffuse emissions of substances that affect human health, especially SO_2, NO_x and CO. Geographical spreading of polluting firms was regarded as a solution. A national measuring network was established. It has been calculated that heightening of industrial chimneys resulted in about 30 per cent decrease in acid deposition in the Netherlands (Zwerver et al. 1984).

Objectives

The policy targets concern both deposition and emission targets. In the NEPP (1989), the interim deposition objective was 4,000

Aeq by 1994 and at the most 2,400 Aeq by the year 2000. After 2000 further reductions should take place. For forests the target is 1,400 Aeq by the year 2000. In order to realise the target for the total deposition, emission targets per acidic compound have been drawn up. The objectives and measures to abate acidification are translated into specific targets for SO_2, NO_x and NH_3.

The SO_2 emissions are mainly produced by refineries (39 per cent), industry (23 per cent), power plants (18 per cent) and traffic (13 per cent). The emission of SO_2 in the Netherlands decreased between 1980 and 1991 by almost 60 per cent. The NEPP-plus objective for the SO_2 emissions by 2000 is a reduction by 85 per cent with respect to 1980, which is likely to be achieved (RIVM 1993). By 1990, the SO_2 emission of refineries and power plants decreased to 35 per cent of the level of 1980, as a result of more stringent emission standards. Industry also achieved important emission reductions.

Traffic is the most important source of NO_x emissions, causing 63 per cent of the total NO_x emissions. Between 1979 and 1991 the NO_x emissions hardly decreased. Cars and power plants became cleaner, but energy consumption and car use increased. The policy target (60 per cent emission reduction from 1980 to 2000) is not likely to be realised.

Over 95 per cent of NH_3 emissions are caused by the agricultural sector. These emissions should decrease by 70 per cent by the year 2000 with respect to 1980. Between 1980 and 1991 the deposition of NH_3 dropped by 17 per cent (Adriaanse 1993), mainly because after 1987 the use of fertilisers (containing nitrogen) diminished. The development of NH_3 emissions depends highly on the developments in the livestock sector. If technical means fail to provide a satisfying solution, the number of livestock has to be reduced.

Effectiveness

Total acid deposition in 1991 was approximately 39 per cent lower than in 1980 (Adriaanse 1993). It is expected that the measures resulting from NEPP policy will reduce the acid deposition to about 2,700 Aeq by the year 2000 (RIVM 1993). The difference between expectation and objective (2,400 Aeq) is partly due to the emissions in surrounding countries which have been reduced to a lesser extent than in the Netherlands. In 1989, over 45 per cent of the average acid deposition in the Netherlands was caused by emissions by foreign countries. It is expected that this percentage will increase to 60 by the year 2000. However, the objective of an average of 1,400 acid equivalents on forest by 2010 will not be realised by current policy. A reduction to about 2,400 acid equivalents may be possible by 2010 (RIVM 1993).

Eutrophication

Eutrophication disturbs ecological processes by an excess of nutrients. Nitrogen, phosphates and potassium compounds are the most important causes of eutrophication. Sources of these nutrients are domestic and industrial waste water, animal manure, fertilisers and deposition from the air. Effects of eutrophication are deterioration of nature and contamination of ground water and surface waters. In a growing number of places, water is not fit for drinking purposes. Ponds and lakes contain undesirably large quantities of algae. Moreover, plant species that thrive in low nutrient environments are disappearing.

The main target group that causes emissions of phosphate and nitrogen compounds is agriculture. In the last 25 years intensive livestock farming has expanded rapidly. Relatively cheap imported feedstuffs were an important input for this industry. Today about 70 per cent of the agricultural area is over-saturated with phosphate and other minerals, caused by an excess supply of manure.

The current policy objective for eutrophication is to restore the balance between the supply and removal of phosphate and nitrogen from water and soil, to safeguard natural processes. Ground water used for drinking must meet the drinking water standards. For agriculture this means that the fields should not be supplied with more nutrients than what the crops can absorb. For discharges by industry and sewage treatment plants into surface water, several specific targets have been drawn up. The Netherlands in accordance with the Rhine and North Sea Agreements was to reduce discharges of phosphate and nitrogen in surface waters by 50 per cent in the period 1985 to 1995. According to the NEPP, by the year 2000 discharges of phosphate must be reduced by 75 per cent and nitrogen by 70 per cent in surface waters with respect to the 1985 level (RIVM 1993).

To abate eutrophication caused by excess manure, standards have been introduced that limit the amount of manure that may be spread over fields. These standards will gradually become stricter by the year 2000 when a definite standard comes into force. This standard will be based on the principle that phosphate and nitrogen input (in the form of manure and fertiliser) should equal output (in the form of phosphates and nitrogen contained in farm products). This means that emissions have to be reduced by 70 to 90 per cent with respect to 1985. In 1987, the government introduced charges on surplus manure, aimed at raising funds for the establishment of manure banks and the development and installation of a manure processing industry. Livestock farmers are obliged to keep 'manure accounts'. Input and output of minerals at the farm level must be in equilibrium by the year 2000 (allowing for some inevitable losses).

Phosphorus and nitrate removal at industrial sources and sewage treatment plants must also lead to a reduction of pollution by these minerals. Nevertheless, the targets for the year 2000 are difficult to achieve, mainly because of the presence of accumulated stocks of minerals in the soil which will continue to contaminate the (ground) water for a long time.

Diffusion of hazardous substances

The 'diffusion' theme deals with a large number of substances which put a burden on or threaten the environment. Several hazardous substances have been identified, the emissions of which urgently need to be reduced. Three main categories can be distinguished: pesticides, 'priority substances' and radioactive substances. The 'priority substances' comprise acidifying and eutrophicating substances, (heavy) metals and metalloids, some organic compounds and other substances like asbestos and particulates. This theme also deals with new substances and genetically modified organisms.

Policy objectives are generally based on the alara-principle, requiring 'as low as reasonably achievable' reductions. The risk posed by the priority substances and radioactive substances must be reduced to a level below the maximum acceptable risk (MAR) by the year 2000. At EU level a 'black list' of substances has been drawn up for which emissions must be reduced to zero. The Netherlands agreed in the Rhine Action Plan and the North Sea

Action Plan to reduce most of the priority substances by 50 to 70 per cent by 1995 with respect to the 1985 level (RIVM 1993). During 1985–90 emissions of priority substances exceeded the MAR level.

About 88 per cent of the hazardous substances dispersed in the Dutch environment originates from agricultural pesticides (Adriaanse 1993). There is a downward trend in pesticide use: in 1991, environmental pollution by agricultural pesticides was 12 per cent lower than in 1980.

The dispersion of radioactive substances showed an increase of 14 per cent between 1980 and 1991. This is mainly due to the increased discharge of radon gas from building materials and a growing use of coal for electricity generation.

The government and the industrial sector agreed on the 'KWS–2000 programme' to reduce environmental damage caused by hydrocarbons. Emissions of volatile organic compounds (VOCs) must be reduced by 60 per cent compared to the 1985 level. Between 1985 and 1990, lead emissions into the air have been reduced by 80 per cent, resulting from the introduction and widespread use of unleaded gasoline.

Waste disposal

The theme 'waste disposal' involves the collection, processing, and destruction of solid and liquid waste flows. It comprises solid waste, nuclear waste, sewage sludge and contaminated soil. Wastes cause environmental problems through the dispersion of harmful substances into the environment when they are dumped or incinerated. It is also becoming difficult in the Netherlands to find enough space to store the growing amount of waste and polluted soil.

Originally, environmental policy was directed at sanitation of contaminated soils and prevention of unacceptable risks. A more integrated approach can be found in the NEPP. Two major objectives are:

- promotion of waste recycling and prevention of waste generation;
- setting up and improving the waste disposal structure and making it 'leak proof'

Waste flow

During the period 1977–90 the total waste flow that had to be combusted, dumped, or discharged remained unaltered: over 20 billion kg. This has to be reduced to 14.5 billion kg by the year 2000 (RIVM 1993). After 1990 the waste flow decreased mainly due to a reduction of building and demolition waste and a decrease of vegetable, fruit and garden waste in household refuse.

Table 5.1
Development of Waste Supply and NEPP-plus Target (in million tonne)

	1986	*1990*	*2000 target*
Recycling	29	30	40
Incineration	3	3	9
Dumping	15	17	5
Discharge	3	3	0

Source: RIVM, 1993.

To solve the waste problem, prevention and recycling of waste are combined in the concept of integrated chain management. The *prevention* objective is a 10 per cent reduction by the year 2000 with respect to the extrapolated trend. Prevention is stimulated by covenants (packaging), environmental management systems within firms and rising costs for incinerating and dumping waste. The objective for *recycling* is to recycle 66 per cent of the waste supply by 2000. In 1990, as much as 55 per cent of the waste supply was recycled. Half the recycled waste originates from industry. For the remaining waste, incineration is preferred to disposal. Therefore, the combustion capacity will be doubled to 6 billion kg by 2000.

Waste water

Waste water originates from households and firms. It contains oxygen depleting and eutrophicating substances, heavy metals and organic micro-pollutants. To prevent discharges of waste water into surface waters, waste water is collected and transported to sewage treatment installations. About 96 per cent of households are connected to the sewerage system.

Soils

It has been estimated that the Dutch soil has been severely contaminated at 110,000 locations, of which only 9 per cent have been identified yet. The government's objective is to reduce severe soil contamination to 'a controllable situation' within one generation. Cost estimates of the entire cleaning operation range from US $25 to 65 billion. To spread costs over time, temporary safety measures have been taken, while cleaning up is postponed. The government wants to charge the polluters for the costs of cleaning up, but it is often difficult or impossible to prove the liability of (former) landowners and users. Voluntary cleaning up by industry only takes place when the contamination causes considerable risks. Sediments have also been contaminated. A start has been made to estimate the amount of sediments that has to be cleaned, the risks and the costs involved.

Disturbance

The quality of the direct residential and living environment is the object of the 'disturbance' theme. Reductions in the levels of noise, stench, risks of calamities and other forms of local environmental disturbances are aimed at. The main sources of these kinds of disturbances are traffic, industry, agriculture and recreation.

Noise nuisance

In 1985, 72 per cent of the Dutch population was adversely affected by noise nuisance, including 30 per cent who suffered severe noise nuisance (RIVM 1993). The policy target for the year 2000 is to stabilise these levels. By 2010, severe nuisance should be reduced to a negligible level. Projected increases in traffic and industrial activities will be compensated by insulation of houses, screening of highways and railways and more stringent requirements for motor vehicles. There is confidence that the objective of stabilisation will be realised (except for aircraft noise), however, the 2010 objective cannot be achieved due to the high level of traffic and transportation. The percentage of people suffering from aircraft noise rose between 1985 and 1990 from 12 to 13, and is expected to increase to about 14 per cent. Industrial noise caused nuisance to 15 per cent of the population. As future developments are uncertain, no prognoses for industrial noise have been made.

Stench nuisance

The number of people in the Netherlands affected by stench rose between 1980 and 1989 from 18 to 22 per cent. The NEPP objective is a reduction to 12 per cent by 2000. However, it is expected that only 16 per cent will be within reach (RIVM 1993). Industry accounts for almost half of the stench nuisance and traffic is responsible for about one-third.

Local air pollution

For a number of substances, maximum concentration levels have been established to prevent health damage caused by local pollution. The number of kilometres of roads that exceeded these levels increased between 1985 and 1989. The reduction of emissions per car has been undone by the increase in car use.

External safety

Information about the risk level in the surroundings of industry, transport and storage of dangerous substances is often lacking. To realise the MAR level[4] and to control dangerous situations, this information gap has to be closed to prevent a shift to 'solutions' that involve even higher but as yet unknown risks. Some 100 industrial plants causing potential risks are required to prepare an external safety report.

Dehydration

Dehydration includes all effects resulting from the decline of ground water level, including effects of drought, mineralisation and changes in partition between seepage and precipitation. In 1984, the considerable decline of (ground) water level was recognised as a nationwide problem. In large parts of the Netherlands, the ground water level has declined. This has been caused by improvements in drainage of agricultural areas and cities, and the extraction of ground water for drinking water production. Over two-thirds of all drinking water is extracted from ground water sources. Industrial and agricultural processes also use large amounts of ground water. As far as the ground water level is concerned, nature and agriculture have opposite interests. While drainage is for the benefit of agriculture, it causes problems for nature.

The policy objective is a decrease of 25 per cent of the dehydrated area by the year 2000 with respect to 1985. This is not likely to be realised. According to the RIVM, a reduction of only 2 to 3 per cent will be achieved by 2000. In spite of measures taken by provinces, municipalities, water management boards and water companies, in 1993 the area of dehydrated lands was still rising. This is due to long procedures, lack of adequate instruments and knowledge of solutions. Large-scale changes in water management and ground water extraction regimes are difficult to realise and have not yet been implemented (RIVM 1993).

Squandering

The 'squandering' theme concerns the depletion of natural resources. Renewable resources should be 'harvested' within the limits of regeneration and depletion of non-renewable resources should be prevented. The NEPP suggests 'integrated life cycle management', energy saving and more attention towards quality, as key elements in the management of natural resources. Policy on this theme is still in its infancy. Integrated life cycle management and the sustainability principle are being studied. In contrast to the output side of social processes (waste flows and emissions), no national objectives have been formulated for the input side (energy, exhaustible resources, nutrients, etc.).

Conclusions

For the various themes of environmental policy, objectives have been formulated and, in many cases, quantified. This makes it possible, in principle, to monitor (lack of) progress and to assess the effectiveness of environmental policy. A downward trend for many emissions and other quantities affecting the environment has been observed. However, most of the objectives for the year 2000 are not likely to be achieved. Therefore, the Dutch environmental policy can be said to be only moderately effective.

Industrial emissions to air and water have been reduced substantially, and progress in this area is still continuing. The main sectors of environmental concern nowadays are agriculture and transport (OECD 1995). As far as industry is concerned, the emphasis is likely to shift from emissions to issues like resource and energy conservation, waste reduction, and 'greener' products.

The successes and shortcomings of Dutch environmental policy cannot be easily attributed to single policy instruments. In general, one can say that extensive regulation in combination with the acceptance of responsibility by industry has played a major role in achieving emission reductions. Negative factors have been a lack of adequate enforcement, the existence of pollution 'heritages' and imports, and the off-setting effect of growth in production and transport. As we have seen in Section 5.2.2, the role of economic instruments has been modest until now. Wherever used, they have been quite effective in contributing to emission reductions.

5.4 COSTS OF ENFORCEMENT AND COMPLIANCE

5.4.1 Introduction

Environmental policy leads to costs for government, firms and individuals. Identifying and quantifying these costs is not an easy task. Clear-cut, consistent definitions as well as operational and standardised methods for measuring the costs are often lacking.

Keeping these limitations in mind, we will provide in this section some data on the costs of pollution control in the Netherlands. At the aggregate level, one can make a distinction between the costs actually incurred and the costs of proposed or planned environmental policy.

The *actual* costs of environmental control are being registered by the Dutch Central Bureau of Statistics (CBS). Their definition includes the cost of all activities which are aimed at protecting, restoring and improving the state of the (external) environment (water, air and soil, including noise[5]). Excluded are items like the costs of refraining (for environmental reasons) from intended activities, cost of environmental damage, and the costs of substitutes for opportunities which have been lost as a result of pollution. Measures to protect occupational health and safety are also excluded.

According to the CBS, total environmental costs in the Netherlands amounted to Dfl. 9.8 billion in 1991 (the most recent year for which this figure is available) (CBS 1994b). Since 1980, these costs have more than doubled. In the OECD (1993b) survey on expenditure on pollution abatement and control, the Netherlands ranks first, together with Germany and the United States. In 1989–90, these countries spent 1.6 per cent of GDP on the environment. They are, somewhat surprisingly, closely followed by the UK with 1.5 per cent. These figures are, however, not readily comparable due to differences in definitions and methodologies across countries.

The *anticipated* costs, which are estimated within the framework of the National Environmental Policy Plans, are not readily comparable with the aforementioned realised costs, because they also contain items like (external) safety and energy conservation. The recent NEPP-2 foresees an increase in costs from Dfl. 17.5 billion in 1995 to Dfl. 22.8 billion in 2000 and Dfl. 26.5 billion in 2010. These costs have been attributed to the various themes, as shown in Table 5.2.

Table 5.2
Costs of Dutch Environmental Policy per Theme (in million Dfl.)

Theme	*1995*	*2000*	*2010*
Climate change	504	664	707
Acidification	2,025	3,234	5,632
Eutrophication	545	1,170	1,102
Diffusion of substances	4,635	5,658	5,875
Waste disposal	7,025	8,640	9,522
Disturbance	852	1,059	1,181
Dehydration	51	164	192
Others (implementation etc.)	1,876	2,170	2,315
Total	17,513	22,759	26,526

The macro-economic consequences of environmental policy are generally moderate to negligible. This is especially the case when the measures are taken on a European level, which prevents environmental costs from leading to a competitive disadvantage for Dutch producers vis-à-vis their European competitors.

Section 5.4.2 deals with the costs of enforcement incurred by public authorities. Section 5.4.3 pays attention to the expenditure by firms on pollution control. Section 5.4.4 gives some conclusions regarding the relative importance of environmental costs in the Netherlands, as well as some remarks on their relationship with the instruments used.

5.4.2 Costs of enforcement

It is almost impossible to get a reliable picture of the costs involved in enforcing environmental policy in the Netherlands. The responsibility for enforcement is divided between several levels of administration (state, provinces, municipalities, water boards), the judicial authorities and the police. The budgets and accounting systems of all these actors do not allow to specify their environmental enforcement costs. Only the budget of the Environment Inspectorate of the Ministry of VROM (which amounted to some Dfl. 17 million in 1994) is spent almost entirely on enforcement related activities. Obviously, this is only a small part of the total costs of enforcement activities in the Netherlands.

A better indication of the enforcement costs can probably be obtained by looking at the number of staff involved in inspections, sample taking and analysing. Table 5.3 gives some data on these activities, as far as they are performed by provincial authorities and the (state and regional) water boards. These figures do not include the enforcement activities by municipalities (which are responsible for most of the smaller firms). Actions based on

Table 5.3
Indicators for the Enforcement of Environmental Law (1991)

	No. of licensed firms	*No. of checks and inspections*	*No. of person–years involved*
Surface waters Pollution Act	32,253	26,382	149
Nuisance Act	5,460	797	11
Noise Nuisance Act	916	949	14
Air Pollution Act	716	1,430	21
Waste Substances Act	1,267	6,530	60
Chemical Waste Substances Act	unknown	917	10
Total	> 40,612	37,005	265

Source: CBS, 1993a, 1993b.

criminal law are also not covered by these figures. Furthermore, one should be aware of the fact that these figures are based on surveys among the competent authorities. The quality of the data may therefore vary, e.g., because of differences in interpretation.

Nevertheless, these figures may be used to assess at least a substantial part of the actual costs of enforcement. Assuming an average cost per person–year of Dfl. 200,000 (including overheads), the costs of these enforcement activities would amount to some Dfl. 53 million per year.

In the NEPP and NEPP-plus, Dfl. 180 million per year was budgeted for support and enforcement tasks. This includes the tasks of lower governments, police and judicial authorities. However, as this amount is not only intended for enforcement, it is probably an overstatement. Without further detailed investigations, we can probably say that the costs of enforcing environmental policy in the Netherlands are in the order of magnitude of Dfl. 100 million per year.

Table 5.3 also gives some information on the extent to which licensed firms are subject to control. Every firm holding a permit under one or more of the six Acts is being checked on average nearly once a year.[6] However, there are large differences in these frequencies. Under the Surface Waters Pollution Act, for instance, 3,770 licence holders were checked one to five times in 1991, 830 were checked six to 40 times and 14 were checked more than 40 times in 1991. Thus, only 4,614 (or 14 per cent) of the total of 32,253 licence holders have been inspected at least once during that year.

In NEPP-2, the government stated as an objective that there should be an adequate level of enforcement by 1995. Lacking concrete targets (e.g., number of checks and inspections), the attainment of this objective could not be verified, although improvements in enforcement can be observed (OECD 1995). Moreover, a structure has been created for a better coordination between the large number of actors involved in enforcement activities.

5.4.3 Compliance costs

The costs of environmental control by industry are being surveyed by the CBS on a yearly basis. These costs are defined as the costs of environmental activities (both performed by the firm itself and contracted), plus environmental levies paid, minus environmental subsidies received. Activities which pay for themselves (for instance, because of savings in raw materials) are excluded. As has been noted in the case study on India (Section 4.4), the environmental costs of progress-integrated (or 'clean') technologies are less easy to identify than the costs of end-of-pipe (or 'clean-up') technologies. In the CBS survey this problem is dealt with by assessing the *additional* costs of the process-integrated technology, compared with an available 'baseline' technology which would have been chosen in the absence of environmental considerations.

Capital costs are calculated on a uniform basis: linear depreciation, based on replacement value and a real interest rate of 5 per cent. The economic lifetime of the construction part of the investment is assumed to be 25 years, and of the mechanical part 10 years.

Table 5.4 shows the costs of industry's own environmental activities, subdivided in capital, operational and other costs. Table 5.5 gives the total annual cost (including charges and payments for activities contracted out) and relates it to some industry figures.

Table 5.4
Costs of Industry's Own Environmental Activities—1989–91 Yearly Average (in million Dfl.)

Industry	*Depreciations*	*Interests*	*Total capital costs*	*Operational costs*	*Other costs*	*Total costs*
Mining	44.5	19.1	63.6	50.4	27.8	141.8
Food	64.1	24.6	88.7	45.9	21.6	156.2
Textiles	2.2	0.9	3.0	1.8	3.9	8.7
Leather and shoes	1.5	0.5	1.9	1.0	0.5	3.4
Paper	10.9	5.0	16.0	8.9	12.6	37.6
Oil	93.9	37.9	131.8	94.7	10.2	236.7
Chemicals and synthetic fibres	184.2	72.9	257.1	233.8	132.4	623.2
Building materials, ceramics and glass	11.1	4.4	15.5	12.1	4.0	31.6
Basic metal	73.4	19.3	92.7	90.1	12.8	195.6
Other metal	30.4	14.1	44.5	30.1	38.9	113.6
Other industries	5.1	2.3	7.3	4.0	9.5	20.8
Public utilities	127.6	51.2	178.8	79.7	48.5	307.0
Total	648.7	252.1	900.8	652.6	322.9	1,876.4

Note: Totals may slightly differ from sums of amounts mentioned due to rounding.
Source: Calculated after CBS (1993c).

Table 5.5
Environmental Costs in Dutch Industry—1989–91 Yearly Average

Industry	*Annual cost (million Dfl.)*	*In % of turnover*	*In % of net value added*	*In % of profit***
Mining	213.4	1.22	1.36*	?
Food	292.3	0.41	1.71	7.6
Textiles	28.9	0.57	1.60	7.5
Leather and shoes	4.1	0.52	1.56	20.5
Paper	69.5	0.82	2.37	9.8
Oil	241.7	1.26	10.97	45.9
Chemicals and synthetic fibres	770.9	1.77	5.81	16.4
Building materials, ceramics and glass	61.6	0.87	2.02	8.3
Basic metal	224.9	2.36	6.46	46.1
Other metals	196.8	0.26	0.74	5.2
Other industries	58.7	0.20	0.53	1.0
Public utilities	458.8	2.34	5.35*	53.6
Total	2,621.6*	0.85	2.17*	10.31

* Based on gross value added.
** For all industries except public utilities, these figures are based on reported profits before taxation. The figures may be overestimated by some 30 per cent, as the profit figures given in CBS (1993e) represent only 76 per cent of Dutch industry (measured by employment).
Source: Own calculations based on CBS (1993c, 1993d, 1993e and 1994a).

From Table 5.5 we can conclude that the largest burden of environmental costs falls on oil, chemical and basic metal industry and public utilities. Together they account for 65 per cent of industrial environmental costs, and for each of these sectors environmental costs are more than 1.25 per cent of turnover, more than 5 per cent of value added, and more than 10 per cent of profits. Except for the leather and shoes industry, in other sectors profit figures were not substantially affected by environmental costs in the period 1989–91. One should keep in mind, however, that these years were relatively prosperous and that the situation may have changed in recent years.

The NEPP-2 foresees substantial increases in the environmental costs of industry, as can be seen in Table 5.6.

The available figures are not detailed enough to predict which industries will get into trouble as a result of increasing environmental control costs. Obviously, the impact will to a large extent

Table 5.6
Actual and Projected Environmental Costs for Industry (in million Dfl.)

	1989–91	*1995*	*2000*	*2010*
Mining	213	212	212	200
Oil industry	242	572	728	689
Public utilities	459	739	1,066	1,200
Other industry	1,708	3,366	4,425	5,080
Total	2,622	4,889	6,431	7,169

Source: Table 5.5, and NEPP-2, Table 7.3.

be determined by other economic factors, such as market developments and the costs of labour and energy.

5.4.4 Conclusions

From the preceding it can be concluded that the Dutch government spends some Dfl. 100 million per year on enforcement of environmental policy, although this figure is surrounded by a large margin of uncertainty. The compliance costs for industry amounted in 1989–91 to Dfl. 2.6 billion per year: less than 1 per cent of turnover, about 2 per cent of value added, or 10 per cent of net profits. The annual compliance costs are likely to double before the end of the century. Differences between industrial sectors are large. The industries which bear the largest burden are dominated by large firms which generally have the capacity to absorb these costs. Nevertheless, the planned environmental policy may have a significant impact on the profitability of certain parts of industry, especially if the economic climate remains harsh for some time. Industry has already expressed its concern over the large costs of the NEPPs, arguing that 'low-cost' options for emission reductions have already been exhausted. They stress the need for international harmonisation of environmental policy, especially within the EU.

In the light of the theme of this book, it is interesting to ask ourselves to what extent the costs of enforcement and control are related to the policy instruments used. However, there has hitherto been no systematic research regarding this question. From the available literature, only some indications can be derived.

The superiority of *economic instruments* over direct regulation, in terms of efficiency, has been claimed frequently in economic literature on theoretical grounds (see Part I of this book). Empirical underpinning of this thesis, however, is scarce. In a review of several studies examining the potential for marketable permits, Tietenberg (1985) found that potential pollution control costs could be reduced by more than 90 per cent in some cases. Wiersma (1989), in a study on SO_2 reduction in the Dutch electricity sector, states that a 92 per cent emission reduction could be achieved at 12 per cent lower costs if an emission charge or tradable permits were applied instead of the usual system of direct regulation.

On the other hand, the cost advantages of economic instruments over direct regulation should not be exaggerated in the present Dutch context. Uniform emission standards are scarce: generally the standards take into account the type and size of the source, its age, etc. In this way, the differences in marginal costs of emission abatement across firms are probably being substantially reduced.

Another way of reducing pollution abatement costs is the use of *voluntary agreements*. A good example is the 'KWS–2000' agreement, in which a general emission reduction target of 50 per cent for volatile organic compounds (VOCs) has been formulated. In the implemention of this plan, this general objective is being translated into differentiated targets for different sources, thus enabling industry itself to look for the most cost effective solutions. The same kind of approach is taken in recent agreements with various branches of industry. In these cases, however, it is clear that only intra-branch differentiations are possible, and not inter-branch ones.

5.5 PROSPECTS

Environmental policy has become firmly established among the main areas of government interference in the Netherlands. The protection of the environment is incorporated in the Constitution (Article 21) as a task for the government. The government declared environmental improvement to be one of the main pillars of its policy. Sustainable development has become a leading concept for all kinds of public and private organisations. Given the fact that the Netherlands is a rich, industrialised country, as well as a small and open economy, the inter-relationships between domestic and global environmental issues are strongly felt. There is considerable concern about the consequences free trade may have for environmental quality abroad. At the same time, initiatives are taken to implement the accepted international environmental responsibilities, e.g., through afforestation and conservation programmes in developing countries, and through environmental codes of conduct by multinational firms.

Although the operationalisation of the concept of sustainable development remains difficult and controversial, Dutch environmental policy has succeeded in formulating mid-term policy objectives for several themes. This creates a reasonably clear perspective for all players in the field ('target groups'). They know what efforts are expected from them, and their performance can be evaluated against these objectives. Obviously, this does not guarantee that the objectives will be reached in time, or even that developments are heading in the right direction. Generally speaking, one could say that the Netherlands is better at making plans and setting objectives than at implementation and enforcement.

Direct regulation has always been the main category of environmental policy instruments, and it will probably continue to be so in the future. Nevertheless, a shift in emphasis is observable. The government wants to take advantage of the 'self-regulating' forces in society which can be promoted by the use of economic and communicative instruments in addition to direct regulation. The exact composition of the instrument mix will vary, depending on the kind of environmental problem and on the specific characteristics of the industry.

Economic instruments like incentive charges and deposit–refund systems will probably be introduced in some specific areas, mainly regarding products which generate waste and/or which are potentially harmful. The use of emission-oriented incentive charges or tradable permits is not very likely. Most actors on the scene are reasonably satisfied with the present system, in which consultations, agreements and permits play the main role. Environment related taxes, such as those on energy, ground water, waste and raw materials may gain some importance. Their rates will probably not become sufficiently high to have a direct influence on supply and demand, but they may play a (small) role in the technological development which economises on the items on which the taxes are based.

The social constraints of introducing substantial incentive changes at the *national* level have been encountered in the discussion on the introduction of energy/carbon taxes in the Netherlands. The discussion focused on the results of calculations of the Central Planning Bureau (CPB 1992) on the impacts of a substantive charge raising energy prices by about 50 per cent in OECD (case A) or in the Netherlands alone (case B). The calculations showed that the impact of such a charge for the Netherlands alone would be quite severe in economic terms: GNP and consumption would drop by about 6 to 7 per cent over the period 2000–2015. Employment would drop sharply by 4 per cent in the year 2000, slightly recovering in the years thereafter. Energy use would drop by 35 per cent: however, the major share of this drop would not be saved, but simply relocated to productive activities outside the national borders. Even an OECD-wide charge would not perform substantially better. Relocation of industries outside the OECD area was assumed to be major.

Although the CPB study has been criticised on several occasions (see, e.g., Opschoor & Turner 1994), its main message has been well understood. The aim of the government is now to stimulate the debate on the introduction of incentive energy/carbon taxes at international levels. But even at the EU level, the impacts on competitiveness of such a charge should be taken into account, possibly by import and export facilities to protect the EU economy against relocative tendencies (Opschoor & Turner 1994). The ideal would be of course to introduce a global tax to counter an essentially global environmental problem.

Communicative instruments will continue to play an important role in (*a*) making people aware of environmental problems and their possible solutions, (*b*) enhancing their own sense of responsibility, (*c*) stimulating environment-friendly initiatives, and (*d*) creating support for environmental policy. The presence of an adequate social basis is considered to be extremely important, both for the general policy objectives and for each individual measure. Lack of such a basis will lead to non-compliance or even sabotage. As a consequence, the sectors in which this 'internalisation' process is the least successful will probably also be confronted with the least stringent policy objectives.

Cooperation has become a keystone of environmental policy, resulting in various agreements between the government and (groups of) firms. The involvement of other departments, firms and environmental organisations indicates how the Dutch environmental policy tries to find a middle course between CAC and let-the-market-do-it positions. Ambitious objectives combined with emphasis on social internalisation and self-activity of the target groups will continue to remain a dominant characteristic of Dutch environmental policy. It still has to be seen whether this 'managed consensus' can realise a development towards sustainability.

Notes

1. Centrale Raad voor de Milieuhygiëne (CRMH), since 1993 called 'Raad voor het Milieubeheer'.
2. Before 1 November 1993: EC (European Community).
3. Preliminary data, CBS, unpublished.
4. This level is defined as the risk of 10^{-6} fatalities per source per year.
5. The costs of nature and landscape protection are being registered separately.
6. One firm can have permits under several Acts at the same time, and one inspection can also be related to several Acts at the same time. However, not all firms which should have a permit actually have one. Under the Surface Water Pollution Act, for instance, less than half of the firms have a permit. In many cases, firms are 'tolerated' because of historical rights, or in anticipation of general rules for the sector to which they belong.

6

CASE STUDIES

The case studies[1] here are based on a comparative appraisal of the working of pollution control measures in three branches of industry—textiles, cement and fertilisers in the Netherlands and India. Though textile industry covers cotton, silk and man-made fibre and fabric industries, our focus is on the cotton textile sector. The character of the pollution problem in each industry is discussed, along with differences between the two countries in this regard. The pollution control standards are compared and the status of their implementation is discussed. As a background, the institutional setting in which the respective industries in the two countries are situated is also presented, so that the problems can be better appreciated.

An issue of central interest in the case studies is to see (*a*) how economic instruments of pollution control work in the two countries in the respective industries, and (*b*) whether there is a case for their greater use, particularly in India. The suitability of economic instruments such as effluent/emission charges and marketable permits would be assessed in the Indian context on the basis of criteria discussed in Part I (2.1). These criteria are: environmental effectiveness, economic efficiency, equity, administrative feasibility and costs, and acceptability.

The three case studies are selected mainly for illustrative purposes. They represent different types of industries as well as different types of pollution problems. The textile industry mainly creates water pollution, cement industry creates air pollution while the fertiliser industry creates both types of pollution.

As a first step, it may be useful to know how these industries stand in relation to the total industrial sector. Table 6.1 presents a

Table 6.1
Textile, Cement and Fertiliser Industries in Relation to the Total Industrial Sector (averages of 3 Years: 1986–87, 1987–88 and 1988–89).

	Textiles	*Cement*	*Fertiliser*	*All industries*
India				
1. No. of units	6,110	585	528	101,543
	(6.0)	(0.6)	(0.5)	(100)
2. Total value of output	119,903	28,271	53,073	1,571,218
Rs million	(7.6)	(1.8)	(3.4)	(100)
3. Total investment	63,997	35,309	54,083	1,148,603
Rs million	(5.6)	(3.1)	(4.7)	(100)
4. Workers employed	942,807	80,169	77,779	7,770,287
(No.)	(12.1)	(1.0)	(1.0)	(100)
5. Profits or loss (−)	−3,515	353	−1,009	44,368
Rs million				
Karnataka				
1. No. of units	278	32	34	5,679
	(4.9)	(0.6)	(0.6)	(100)
2. Total value of output	2,780	2,422	1,316	66,665
Rs million	(4.2)	(3.6)	(2.0)	(100)
3. Total investment	1,959	3,110	2,118	48,978
Rs million	(4.0)	(6.3)	(4.3)	(100)
4. Workers employed	29,975	8,199	2,025	397,970
(No.)	(7.5)	(2.1)	(0.5)	(100)
5. Profits or loss (−)	−468	227	−277	493
Rs million				
The Netherlands				
1. No. of units	80	29	9	5,831
	(1.4)	(0.5)	(0.2)	(100)
2. Total value of output	18,949	10,118	23,178	2,540,950
Rs million	(0.7)	(0.4)	(0.9)	(100)
3. Total investment	25,687	19,981	32,654	2,810,890
Rs million	(0.9)	(0.7)	(1.2)	(100)
4. Workers employed	9,743	3,822	4,830	760,310
(No.)	(1.3)	(0.5)	(0.6)	
5. Profit or loss (−)	1,423	1,776	67	157,483
Rs million				

Note: Figures for India include only the registered factories and exclude household and small-scale sector relevant in the case of textiles. Registered factories are those employing 10 or more workers using power, or 20 or more workers without using power. There is only one fertiliser manufacturing industry in Karnataka, but there are others which are formulating units. Moreover, 'fertiliser' industry is clubbed together with pesticide industry in ASI data taken here.

Table 6.1 (contd)

Figures for the Netherlands are average values of 1987–89. Included are firms with 20 or more employees in the following branches (SBI codes): 22.1, 22.2 and 22.4 (textiles); 32.4 and 32.7 (cement); 29.1 (fertiliser); and 2 and 3 (all industries). Profits are those before taxation. Investments is the value of the stock of capital goods. This value has been calculated by assuming a uniform capital:output ratio for the entire industry to which the branch belongs (capital stock statistics were only available at higher levels of aggregation). Dutch guilders have been multiplied by 10 to arrive at rupee figures.

Figures in brackets are percentages to 'all industries'.

Source: Annual Survey of Industries: 1988–89, Government of India, New Delhi, 1992. Figures for the Netherlands are based on Central Bureau of Statistics (several publications).

brief overview. Of the three industries, textiles are far more important in India than the other two in all respects. Textiles are an important source of foreign exchange earnings, accounting for about 30 per cent of total exports. About 20 per cent of the turnover of the industry is in the form of export earnings. On the other hand, cement and fertilisers produced are used for domestic consumption. In the Netherlands, the industries are more equal in size. Fertiliser is bigger in turnover and investment while textiles employ more workers.

In relation to the total registered factory sector in India, the textiles account for 7.6 per cent of value of output and 12.1 per cent of employment. In Karnataka, they are relatively less important, but nevertheless significant in the industrial map of the state. Cement and fertiliser are more capital intensive than textiles, as can be seen from Table 6.1.

From the point of view of capacity for pollution generation, all the three industries are important, more than proportionately to their share in output and investment. During 1980–81 to 1990–91, the decadal growth was as high as 207.7 per cent in the case of synthetic textiles in Karnataka, and 202.2 per cent in the case of cement. In contrast, the cotton textile and fertiliser sectors recorded a moderate decadal growth of 62.5 per cent and 66.8 per cent respectively. These are modest rates of growth also compared to the decadal growth of total industrial production in the state during the same period, which was 162.5 per cent.

From the point of view of profitability, textile and fertiliser industries are not in a good shape. Both at the all-India and the state level, the two industries have made losses though there are individual factories which have made profits. Since we have used the average figures for three years, 1986–87 to 1988–89, the losses cannot be attributed to abnormality of a given year. Instead they reflect an enduring problem of the industries as a whole. On the other hand, the cement industry showed profits both at the national and state levels (respectively 1.0 and 7.3 per cent over investment, and 1.2 and 9.4 over the value of output). Interestingly, the profitability of cement industry in Karnataka is much higher than at the national level.

In the Netherlands, the relative importance of the three industries considered is less than in India. Profitability in the textile and cement industries was satisfactory in 1987–89. The fertiliser industry, on the other hand, suffered losses in two of these three years.

6.1 TEXTILE INDUSTRY

Textile industry is far more heterogeneous and complex than cement and fertiliser, particularly in India. It includes spinning, weaving, finishing and sewing sectors.[2] From the point of view of pollution, only the finishing sector, comprising bleaching, dyeing and printing, is relevant in the case of cotton as it alone generates water pollution. While it is easier to separate the finishing sector from other sectors in the Dutch case due to a greater degree of specialisation, this is not so in the Indian case. On the contrary, it is more meaningful to classify the Indian textile industry into three categories: handlooms, power looms and the mill or factory sector. The mill sector has mills specialising in spinning and weaving or in finishing, and also composite mills. The first two constitute the unorganised small sector while the latter comes under the organised sector. The handlooms and power looms often make use of independent processing houses for finishing within the unorganised sector or outside, while the composite mills have their own in-house processing facilities. The organised sector is in an unfavourable position compared to the other two. The Factories Act with all its restrictions and protective clauses for labour is not applicable to the unorganised sector. It evades not only the labour laws, but also excise duties and pollution standards. The unorganised sector is promoted mainly to generate employment opportunities in a decentralised way in far-flung villages and small towns. The power looms particularly offer tough and unfair competition to the mill sector. While handlooms and power looms engaged only in spinning and weaving are not a source of pollution, the independent processing houses in the unorganised sector patronised by them are a source of water pollution and manage to evade anti-pollution measures. However, they have not assumed the same menacing proportions as tanneries which are also in the unorganised sector, and have not yet raised a public outcry as in the case of tanneries. The pollution caused by the unorganised textile industry has not invited much attention of the PCBs.

Both the handloom and mill sector enjoy government support. Handlooms are given direct subsidies because of the employment creating potential. While the mill sector as a whole does not receive government subsidy and support, there are a significant number of units which are declared sick or economically not viable, but are continued to be operated under government-owned National Textile Corporation or State Textile Corporations. This is done again because of large-scale unemployment following the close down of the mills. They suffer from excess capacity, outdated technology, shortage of funds and stagnant demand conditions (see *Facts for You*, March 1992). This is not characteristic of the whole of the mill sector, but certainly of a significant part. It is to be appreciated that even in these circumstances, the textile industry is expected to observe pollution control standards.

The Dutch textile industry, fortunately, does not have such disadvantages. There are 80 textile factories in the Netherlands, employing nearly 10,000 workers.[3] It is a far less important sector in the Netherlands than in India, contributing 0.7 per cent to total industrial turnover and affecting 1.3 per cent of industrial employment. About 45 per cent of its turnover is in foreign earnings (Amelrooy 1994, p. 48). Though the industry faces competition from countries with cheap labour, this is more than compensated by high level of automation and modernisation.

Pollution

The pollution problem caused by the textile industry is complex. The effluents contain not only organic substances, but also several substances which are hardly degradable. The PCBs in India have specified almost 30 parameters to be monitored, of which only about five most important are monitored. The pollutants cannot be abated by a single stretch of treatment. It requires primary, secondary and tertiary treatment plants involving considerable expenditure. The standard procedure of effluent treatment includes a neutralisation tank to lower pH value by adding alkalis or acids, an aeration tank to bring down BOD and COD levels, and a clarification tank to settle down solids with the help of alum. Finally the treated effluents are let out into a sewerage or an open stream. The lower part consisting of sludge is dried and disposed. The presence of heavy metals (which are hardly removed) prevents it from being freely used as organic manure.

In the Netherlands, the Surface Water Pollution Charge (SWPC) was introduced in 1970 to finance municipal sewage purification plants. Though intended to be mainly a fund-raising instrument, it began to have a regulatory impact. Till about 1970, industrial production and organic industrial waste water together increased in the country, but thereafter, while industrial production continued to increase, organic waste water declined sharply.

Two components of water pollution are distinguished: oxidisable or organic substances and the non-oxidisable substances like heavy metals. Both are taken into account in the charge base, and units of pollution (PUs) are defined accordingly. As for oxidisable substances, one PU of households' waste water equals 180 grams (since 1986: 136 grams) of oxygen demand. The pollution load of industrial waste water is found by dividing the oxygen consumption by 180 (136). The organic pollution load (in PU) of industrial waste water is deduced from the pollution load of one household's waste water. For the non-oxidisable substances, the units of weight discharged per time in the surface water determine the pollution load. One kg discharged per year of these substances is counted as one PU, except for cadmium and mercury. For them, 100 grams per year equals one PU. The pollution load of a company consists of various pollutants, and their total load in PU is the sum of all the pollution units, allowing for a correction for the volume of pollution (Amelrooy 1994, p. 47).

It would be interesting to see how the pollution charge worked in two case studies of textile industry in the Netherlands (Amelrooy 1994, pp. 49–52). One of them is Koninklijke Textielveredeling (KTV), established as early as 1924. The production consists of printing and improving cottons and blends. It employs around 230 people and had a turnover of Dfl. 24 million in 1991, with a realised profit after tax between Dfl. 850,000 and one million. KTV discharges 8,000 units of pollution every year. Every 24 hours, a sample is taken, analysed by an independent laboratory. The formula to determine PU includes water consumption as well, which is 430,000 m^3 per year for KTV. The tariff of a PU is Dfl. 54, resulting in a charge burden of around Dfl. 430,000. This amounts to a mere 1.8 per cent of turnover. The tariff per PU differs regionally as the costs of the Water Quality Manager are divided by the number of industries and households in the region. Both SWPC and the burden of energy costs contributed to a reduction in the amount of waste water. SWPC also led to a replacement of dyestuffs containing heavy metals. Dyes based on cadmium, chromium and lead have been taken out of production completely. Oxidisation is reported to be done using sugar instead of chromium, or has been abolished completely. Since irregular discharging to a public purification plant led to a heavy fine, KTV also established a buffer tank in 1974, resulting in a more regular discharge. The temperature and pH are equalised as well. It is to be noted that SWPC did not lead to the establishment of a private waste water treatment plant, but only to a buffer tank to level the discharge to the public purification plant. The proceeds of the charge, however, helped to finance the latter.

Garenveredeling Aalten (GA) is a small-scale plant employing only 35 persons, with a total turnover of Dfl. 5 million and realised profits after tax being Dfl. 200,000 in 1991. The pollution load consists of 1,400 PU per year. Sampling is done twice a year during one week continuously. Its water consumption is 125,000 m^3 per year. The charge rate is the same as for KTV (Dfl. 54 per PU) as both are in the same region. The total charge is about Dfl. 75,000 per year which amounts to a mere 1.5 per cent of turnover. The factory does not find it economical to treat its own sewage, but considers it cheaper to pay the charge. But the charge has induced the company to economise on water. Since 1995, however, it has become compulsory to have a private treatment plant, and the dyes have to be purified from waste water up to 90 per cent. The GA says it uses only bio-degradable dyes which do not affect the environment but do colour the water. Since dyes are soluble, they are difficult to remove from water.

Though both these textile units have not considered it economical to have their own private waste water treatment plants, the pollution charge cannot be considered as a failure. It has induced them to change over to dyes that are less harmful to the environment. It has also induced them to keep the pollution load to the minimum. Moreover, the charge has helped to raise funds to finance public purification plants. The charge has given freedom to factories to decide whether or not to have their own treatment plants, resulting in more efficient and cost effective anti-pollution measures. Firms which find it cheaper to abate pollution themselves can do so, while others will pay the charge which can be used to abate pollution under the public sector. The general idea is that somebody has to abate pollution, who will do it has to depend on who is most cost effective. An interesting fact which comes out from the Dutch experience is that the pollution charge has not replaced regulatory measures. It has only supplemented them in an important way. If pollution abatement by the public sector is uneconomical, regulatory measures can still insist on factories having their own treatment plants. Regulation, therefore, is a more dominant measure compared to the pollution charge.

It is to be noted, however, that there are no legally obligatory Minimum National Standards (like MINAS of India) in the Netherlands. The standards are laid down in individual permits issued by the Water Quality Manager (WQM) to respective industrial units. The WQM is free to set the standards, subject to certain conditions. For example, substances belonging to the black list consisting of very toxic, persistent and bio-accumulating substances have to be controlled with the best existing technologies. There is no standard per substance but an obligation to make every necessary effort to abate pollution. In India, the SPCBs can prescribe more stringent standards than MINAS, but cannot relax them.

This is in theory. Let us see with the help of actual examples how far pollution standards are followed and what difficulties were reported in textile firms in this regard in India. Since the implementation of standards gives a mixed picture, it would be instructive to study case by case. In all, seven textile units were surveyed by us, one of them being a small-scale processing unit.

There is little to tell about the small-scale unit located in Bangalore engaged in dyeing silk fabrics. It is simply unaware of pollution standards and lets out effluents into the municipal sewerage system. Unlike the Dutch small-scale unit, it pays no charge and its pollution load is not monitored. On the other hand, the other units which are large or medium scale, have installed Effluent Treatment Plants (ETPs), but often their capacity is not adequate to treat the entire waste water generated. Out of nearly 30 parameters which in principle should be monitored for waste water according to standards, only pH, BOD, COD, TDS (Total Dissolved Solids) and TSS (Total Suspended Solids) are monitored. The discharge of heavy metals is not monitored, but they were reported to be not in use.

The case of Minerva Mills, a large-scale unit, can be taken to be illustrative. It reported results of a random inspection as shown in Table 6.2.

Table 6.2
Pollutants vis-à-vis Standards in the Case of a Textile Mill

Constituents	*Obtained*	*Standard*
pH	8	6.5 to 8.5
COD	199 mg/l	250 mg/l
BOD	18 mg/l	30 mg/l
TDS	3,700 mg/l	2,100 mg/l
TSS	148 mg/l	100 mg/l

Source: KSPCB (personally gathered in 1992) and from the response to questionnaire by the mill.

The standards are met in respect of the main constituents—the first three. TDS and TSS do not meet these standards and depend on the public purification plants. Minerva Mills argues it is not economical to meet these standards with available technology and that the standards are too stringent. The mill has a turnover of Rs 201 million, and the annualised cost of pollution control is only Rs 0.46 million, forming a mere 0.23 per cent of turnover. The cost burden on account of pollution control for this and other textile units is presented in Table 6.3. Minerva Mills is a sick unit under NTC and making losses. It is creditable that in spite of losses, it has undertaken pollution control by setting up ETP and operating it. During 1991–92, it made a loss of Rs 30.5 million, which would not have materially reduced even if it had not incurred expenses on pollution control.

There is a similar problem with Srinivasa Mills. Established in 1972, it had an ETP with a capacity for 200 kl/day. Since then its production has increased and waste water flow has increased to 500 kl/day. As such, it is not able to meet BOD standards. The

Table 6.3
Cost Burden on Account of Pollution Control in Textile Industry (average of 3 Years: 1989–90 to 1991–92)

Industrial unit	*AC on PC (Rs in million)*	*Total AC*	*Sales*	*Proportion (%) of AC on PC to:*		*AC on PC per thousand litres of waste water treated—Rs*
				Total AC	*Sales*	
Minerva	0.46	32.81	201.00	1.4	0.23	13
Srinivasa	0.36	4.75	51.44	7.6	0.70	3
Gemini	1.24	32.12	140.00	3.9	1.0	7
Binny	0.36	32.48	563.70	1.1	0.1	1
Davangere	0.43	93.55	360.25	0.5	0.1	1
Gogte	0.33	35.61	90.64	0.9	0.1	3
Average	0.53	38.55	234.50	1.4	0.23	2

AC: Annualised Costs
PC: Pollution Control
Source: Questionnaires canvassed with respective units.

KSPCB has filed a suit against the mill and renewal of its consent is pending. The factory has made known its intention to acquire ETP to process the present level of waste water and is allowed to function normally.

Since Srinivasa Mills is not located in Bangalore, it cannot release its waste water into the public sewerage, and has to treat its own waste. Being smaller than other units studied, its pollution control costs are higher than those of others, especially in relation to their total costs (see Table 6.3). As a proportion to turnover, these costs were marginal (only 0.7 per cent). Incidentally, this is the only profit making unit amongst those studied here, but as a proportion of its turnover it was only 1.1 per cent during 1991–92.

Gemini Dyeing and Printing is a large-scale processing unit established in 1989. It has set up an ETP with an investment of Rs 4.5 million to process waste water up to 600 kl/day. It was reported, however, that they were not regularly operating the plant due to power cuts and shortages and have not been able to meet the standards fully. Its annualised costs on pollution control as a proportion of turnover were 1 per cent, which was the highest among the units studied. However, this is still a marginal proportion. Being in a highly competitive field, it is unable to pass on the costs freely to its customers.

Binny Mills in Bangalore is one of the oldest mills in the country, established in 1884. It is unfortunately a sick unit making losses continuously. Nevertheless, it has set up an ETP with a capacity of 200 kl/day. Though the company reported that it is meeting the pollution standards, it was refuted by the KSPCB which held that the company was letting untreated waste water into public sewerage. The renewal of consent was under consideration at the time of our study, but the mill is in operation as its closure would cause unemployment to nearly 3,000 workers. To the extent it is incurring pollution control expenditure, its cost is only 1.1 per cent of total annualised costs, and only 0.1 per cent of turnover. Even if it were to meet the monitored standards fully, the increased costs thereby would be well within marginal limits. But since it is a loss making and an old mill, the matter is complex. The mill nevertheless has plans to recycle the waste water after treatment.

Davangere Cotton Mill is located in Davangere (about 230 km northwest of Bangalore) and is among the oldest in the state being established in 1939. It has also been making losses continuously. Nevertheless, it has set up ETP and was reported to be meeting all the monitored standards. This judgement is based not only on the statement by the industrial unit concerned but also on the basis of information cross-checked with KSPCB. The factory has its own in-house laboratory, and samples of water are tested regularly. One of the reasons why it is meeting standards while others are not could be that it is close to two controversial polyfibre units at Harihar which attract a lot of public attention. The local people are alert about any untreated effluents discharged into open streams. The Davangere Mill discharges treated effluents into the municipal sewerage system. The costs incurred by it on account of pollution control are negligible in relation to its total costs and sales (see Table 6.3).

Gogte Textiles is a composite mill located in Belgaum, some 460 km to the northwest of Bangalore in Karnataka. It is a relatively recent mill, established in 1980. It has its own ETP plant and has been meeting the monitored standards both as per the company and KSPCB. The treated effluents are discharged within premises for agricultural purposes. They do not have their own laboratory but the effluents are tested by authorised laboratories. They propose to have their own laboratory and separate pollution control staff soon. It is also a loss making unit (Rs 43.8 million loss during 1991–92, which was as high as 45.5 per cent of their gross sales [turnover]). But this has not come in the way of pollution treatment. The costs arising from pollution control are very small in relation to total costs and sales (see Table 6.3).

No account of pollution problems caused by the textile industry in Karnataka is complete without referring to two Birla units located in Kumarpatnam near Harihar, about 250 km northwest of Bangalore, on the banks of Tungabhadra river. Harihar Polyfibres Limited (HPF) was established in 1972 to manufacture rayon grade pulp based on a mixture of eucalyptus and some tropical hardwoods. In 1977, the Gwalior Rayon Silk Manufacturing Company (GRASIM) started producing 'grasiline' fibre using rayon grade pulp manufactured by HPF. These two factories discharge their treated effluents into the river. The effluents from HPF have conformed to standards while the effluents from GRASIM contained zinc, total dissolved solids and sulphates in excess of standards laid down (*Jagruta Vani*, Vol. 7, Nos. 2, 3 and 4 combined, July–December 1991, pp. 23–24). As a result, aquatic life was reported to be adversely affected for a stretch of about 500 metres. A major fish kill in 1984 spurred a popular movement against pollution by the factories, and the two units have been in the centre of public controversy ever since. (We could not succeed in getting a response to our questionnaire nor an interview and hence are unable to give a picture regarding their investments and expenditure on pollution control.) HPF is reported to have developed a new bleaching technology based on oxygen, by which they were able to do away with the requirement of conventional chemicals. This has reduced the costs as well as pollution. For this technology development, the unit has won an award from the Indian Chemical Manufacturers' Association in 1991 (*The Financial Express*, Bombay, 14 September 1991).

We have noted the cost burden on respective firms due to pollution control, in terms of total costs and sales in Table 6.3. The last column in the table presents annualised costs on pollution control per thousand litres of waste water treated. Interestingly, though the largest two firms had the lowest cost of treatment from this point of view (as also in terms of sales), the smallest among firms did not have the highest cost. The size of the firm showed no strong correlation with the cost levels. Since the number of our observations is small, we have not calculated the correlation coefficient. The point, however, is that though in terms of weighted average, the cost of treatment within textile units was as low as Rs 2 per kl, it varied from Re 1 to Rs 13 within the five units selected.[4]

The review of some of the large- and medium-scale textile units showed that all of them have set up ETPs and are operating them, but not all the standards are met. It is heartening that even loss making units have set up ETPs and some of them like Davangere and Gogte Mills are meeting all the monitored standards. In terms of costs, setting up ETPs and operating them does not make a big difference to profits or loss, as they account for a very small proportion of total costs and sales. This is a finding of some significance to us while considering the feasibility of economic instruments like effluent charges.

Since a significant part of the output of textile industry is exported (about 20 per cent), the industry is coming under pressure to observe environmental standards particularly in respect of bleaching and dyeing. Traditionally, Indian textiles were using natural dyes which were eco-friendly, but they were replaced by synthetic dyes as the latter were amenable to greater standardisation and imparted fast colours. Attention has been turned again towards natural dyes, and even problems of fastness are being overcome (Chandramouli et al. 1993; Venugopal 1993a and 1993b). Europe and USA have banned the use of benzidine and other carcinogenic chemicals. Synthetic dyes and chemicals containing chlorine or chlorides including bleaching powder are actively discouraged. The Indian textile industry, therefore, has no option but to be eco-conscious and eco-friendly even in respect of dyes and other chemicals being used, if it is to continue to export. This is bound to desirably affect even production processes for domestic industry.

6.2 CEMENT INDUSTRY

In India, the cement industry made a modest beginning in 1914 with the establishment of Indian Cement Co. Ltd. in Porbander

(Gujarat) with an installed capacity of 1,000 tonnes per year. At the start of planned development in the country in 1950–51, India produced 2.2 million tonnes, but by 1992, the industry reached a production figure of 54 million tonnes. But this was not a smooth growth. There were price and distribution regulations up to 1982 which acted as a disincentive to production. After a partial decontrol was announced that year, production jumped from around 18 million tonnes to 46 million tonnes in 1990–91. A total decontrol was announced in March 1989.

Another factor which has affected the fortune of the cement industry is the fluctuations in the level of government demand. A decline in this demand is reported after 1991, as a result of which prices fell from their earlier dizzy heights to more reasonable levels corresponding to costs. However demand prospects for the cement industry are bright on the whole, as the construction sector is rapidly expanding. An indication of further prospects for demand comes from the fact that per capita cement consumption in India was only 47 kg in 1987 as compared to 210 kg in the world as a whole (*Kothari's Year Book 1991*, p. A–82). Understandably, the cement production in India is meant mainly for domestic consumption, hardly 5 per cent of cement output is being exported. The exports of cement, however, have been showing a rising trend, but account for a negligible fraction of the total export earnings of the country. There are no imports of cement into India.

Though not as heterogeneous as the textile industry, the cement industry has significant variation in the size of the factories. The government encouraged the promotion of mini cement plants producing less than 200 or 300 tonnes per day, to take advantage of scattered and small reserves of limestone and also to encourage small entrepreneurs and discourage monopoly. But they accounted for only about 7 per cent of the total output in 1990–91 (*Facts for You*, Annual, 1992). They have not much attracted the attention of the Pollution Control Boards. They find it hard to compete with large-scale units, and some of them are taken over by the latter. The capacity of cement factories (other than mini plants) varies from 3,000 tonnes to 50,000 tonnes per day. The average rate of capacity utilisation is about 83 per cent (*Facts for You*, July 1991, pp. 27–29).

Cement industry tends to get concentrated in areas of large reserves of limestone. It is located mostly in the states of Madhya

Pradesh, Tamil Nadu, Andhra Pradesh, Rajasthan, Gujarat, Bihar and Karnataka. Karnataka accounts for less than 10 per cent of the country's cement production. Of the 110 major cement factories (excluding mini plants), 10 are in Karnataka. Most cement units are concentrated in the northern districts of the state—Gulbarga, Bidar, Bijapur and Belgaum.

With decontrol, there was also another welcome and distinct trend towards modernisation of the technology. Since about 1988, there has been a shift from the earlier wet to dry process technology,[5] and even to a more modern precalcination technology. The dry and precalcination technologies are fuel saving and more efficient. The higher investment needs of modernisation could be met through a flow of funds following a sharp increase in the prices of cement stock in the share markets. This enabled quite a few cement units to set up pollution control plants as well.

The main air pollution problem involved in cement production is the emission of particulates and dust. Gases discharged through kiln stacks of the cement industry are not supposed to be harmful to humans, animals and vegetation, but particulates and dust emitted are harmful. They can lead to respiratory diseases in humans as well as in animals, and agricultural yields are reported to decline as a result of these emissions. There are studies to show that the environmental costs so imposed are considerably higher than the costs of pollution control in the case of cement industry (Abubacker 1990). There are several methods of reducing emissions, which involve the installation of dry mechanical collectors, wet collectors, fabric or bag collector, electrostatic precipitators, or pulse dust collectors (Joshi & Chouhan 1991).

The emission standards for particulates and dust laid down in India for cement industry are:

Capacity [TPD: Tonnes per day]	*Protected area*	*Other areas*
200 or less	250 mg/nm^3	400 mg/nm^3
More than 200	150 mg/nm^3	250 mg/nm^3

nm^3: normal cubic metre
Source: EPA Notification, 16 April 1987.

To comply with the standards, most of the Indian cement factories have installed electronic precipitators. A few plants have also decided to install fabric collectors, the filter cloth for which is imported. A few large-scale units, which are financially sound, have installed ultra-modern pulsegen (hot) ESP (Banerjee 1991).

Before we go into the pollution control status of the Indian cement industry, it would be useful to have a background of the Dutch cement industry. The cement industry is dominated by one company in the Netherlands—ENCI Netherlands, which has three factories in the country.[6] The Maastricht plant is the only one engaged in quarrying limestone and converting it into clinker. It provides clinker to the other two factories. The company meets 60 per cent of the demand in the country, the rest is imported from Germany and Belgium (Amelrooy 1994, p. 59).

A problem with ENCI Netherlands is that it is a part of a larger holding company with cement factories in other countries. If pollution control costs become totally out of line in the Netherlands, the production can be shifted to other countries with less stringent environmental regulations. However, this cannot be the only factor, since investments sunk in fixed capital cannot be shifted without cost, and the transport costs may more than outweigh the saving in pollution control costs. The only competition feared is from Belgian cement produced nearby. Since comparable environmental standards apply in Germany, adverse competition from this country is not feared (Amelrooy 1994, p. 61).

Three strategies which reduce costs and are also presented as 'environment-friendly' are tried by the company: (*a*) Produce cement with materials other than clinker. Clinker is replaced by residues as fly ash and slag, which economises on scarce limestone and on energy; (*b*) transition of combustibles from fossil fuels to industrial residues, thus reducing overall CO_2 emissions; and (*c*) installation of new equipment to save on electricity. In 1991, a voluntary agreement was concluded with the government for reducing energy consumption and CO_2 emissions by 20 per cent (Amelrooy 1994, p. 60).

Apart from these strategies, the company is bound to the Dutch emission directives. There are no charges on pollutants as in the case of water pollution, but the directives are far more demanding in the Netherlands than in India.

> At ENCI 50 milligram of dust is emitted per cubic metre from the kiln, while 25 mg is emitted from newer plants. With the investments in electronic precipitators, the emission is brought

> down to only 15 mg. The Dutch directive for emission of dust from large sources[7] is 10 mg/m³. This is a general directive for all industries and it presumes the use of a fabric filter. These fabric filters have the best efficiency. If for some reason it is not possible to use fabric filter the standard is relaxed to 25 mg dust per cubic metre. Such an exception is made for the cement kilns: because of the high temperatures, it is not possible to use fabric filters. However, at the other stages of production there is no need to make an exception; for the crushing, grinding, packing etc. processes, the general directive of 10 mg/m³ applies. If it is economically not acceptable to require an emission level of 25 mg/m³ because of booming costs, the limit is relaxed to 50 mg/m³. The directives are thus not as fixed as standards and do not have the same legal status (Amelrooy 1994, p. 60).

In the Dutch case, it is reported that the environmental costs increased relative to other production costs. 'The environmental costs increase with 10 to 15 per cent, while the costs of raw materials, personnel and maintenance increase with the normal inflation rate of 5 per cent' (Amelrooy 1994, p. 61).

We may now see how far pollution control measures are in operation in India, taking only selected firms in Karnataka as illustrations. The status of implementation in the industry in the country as a whole and also in Karnataka is presented in Table 6.4.

All the three cement units selected in Karnataka have adopted pollution control measures and are satisfying the standards laid down. All of them have electrostatic precipitators and dust collectors while two of them had even imported filter bags in use. They did however report considerable difficulties on account of irregular power supply and sudden interruptions therein. The cost incurred on pollution control measures in relation to total costs and sales is presented for the three units in Table 6.5. Though in the marginal range, they are noticeable as in the case of Kanoria which reported abnormally low sales in relation to costs.

Interestingly, two units reported losses during 1991–92 and one unit showed losses during all the years (1989–90 to 1991–92), and yet they have all implemented pollution control measures. The recoveries of cement dust have an economic value and exceeded the annualised costs of pollution control in the case of two units—Mysore Cements and Kanoria. The concerned data were not

Table 6.4
Status of Pollution Control in Cement Industry

	India 1991	*Karnataka 1993*
1. Total no. of units	94	10
2. No. of units with pollution control in operation	60	8
3. No. of units with pollution control measure in the process of implementation	23	1
4. Not implemented	5	–
5. Units closed	6	1

Source: Assocham 1991, p. 19, KSPCB 1993, p. 45.

Table 6.5
Cost Burden on Account of Pollution Control in Cement Industry
(average of 3 Years: 1989–90 to 1991–92).

Industrial	*Annualised costs*		*Sales*	*Proportion (%) of annualised costs on pollution control to*	
	Pollution control	*Total*			
	(Rs in million)			*Total annualised costs*	*Sales*
Rajashree	15.86	639.36	1,340.96	2.5	1.2
Mysore	11.20	439.22	602.38	2.6	1.9
Kanoria	6.00	136.74	55.35	4.4	10.8
Average for 3 units	11.02	405.11	666.23	2.7	1.6

Source: Questionnaires canvassed.

available from Rajashree. They however reported that the recoveries could be economically significant. In contrast, in the Dutch case the recoveries were reported to be of no economic value as they comprised a very small proportion of production. Such recoveries were not reported in the textile industry.

Another difference between the textile and cement industries is that while in the former, there were hardly any units which maintained separate staff for pollution control and monitoring, all the three cement units reported maintaining such staff consisting usually of one or two managers or supervisors and one or two technicians.

6.3 FERTILISER INDUSTRY

The fertiliser industry made a nominal entry in India with the establishment of a superphosphate factory in Ranipet, Tamil Nadu in 1906 with a capacity of 6,400 tonnes of phosphate (P_2O_5) per annum. In the nitrogen sector, TISCO made a notional start by producing ammonium sulphate $(NH_4)_2$ SO_4 as a by-product in 1933. It was only after Independence that a further boost came with the establishment of ammonium sulphate plant at Sindri in 1951, which coincided with the commencement of planned development in the country (*Kothari's Industrial Directory of India* 1992, p. 5.11).

Only nitrogen and susperphosphate are being produced in India, and all the requirements of potassic fertilisers are met through imports. The figures for production, consumption and imports of chemical fertilisers from 1951–52 to 1991–92 are given in Table 6.6a for India and in Table 6.6b for the Netherlands. India is not self-sufficient in chemical fertilisers, and deficits are met by imports. There is, however, a trend towards increasing self-reliance. Consumption shot up from a mere 70,000 tonnes in 1951–52 to 338,000 tonnes in 1960–61, and as a result of the green revolution, further to 2,687,000 tonnes in 1971–72. In 1951–52, 55.7 per cent of consumption was met by domestic production, and 64.8 per cent in 1961–62. The sharp increase in consumption during the 1960s could not be met by domestic production alone, which met only 46.1 per cent of consumption in 1971–72. But thereafter, the proportion of output to total consumption increased to 67.5 per cent in 1981–82 and 77.5 per cent in 1991–92. This occurred even when consumption shot up from 2.7 million tonnes in 1971–72 to 12.7 million tonnes in 1991–92.

In spite of this phenomenal increase, the consumption of chemical fertiliser per hectare of arable land[8] in India lags behind the world average but the difference is being narrowed down fast. While this figure was 13.2 kg in 1970–71 for India, the world average was 47.4 kg. By 1990–91, the world average had increased to 95.2 kg, while the Indian consumption shot up to 74.3 kg, thus considerably reducing the difference. Karnataka lags a little behind with 67.8 kg of fertiliser consumption per hectare of arable land.

Table 6.6a
Production, Consumption and Imports of Chemical Fertilisers in India
(thousand tonnes)

	1951–52	*1961–62*	*1971–72*	*1981–82*	*1991–92*
Output					
N	29	154	949	3,143	7,302
P_2O_5	10	65	290	950	2,562
Total	39	219	1,239	4,093	9,864
Consumption					
N	55	250	1,798	4,069	8,046
P_2O_5	9	60	588	1,323	3,321
K_2O	6	28	301	676	1,361
Total	70	338	2,687	6,068	12,728
Imports					
N	29	307	481	1,055	566
P_2O_5	16	–	248	343	968
K_2O	8	75	268	644	1,236
Total	53	382	997	2,042	2,770

Source: Fertiliser Statistics, 1971–72, 1981–82 and 1991–92, The Fertiliser Association of India, New Delhi, 1971–82 and 1992.

Table 6.6b
Production, Consumption and Imports of Chemical Fertilisers in the Netherlands
(thousand tonnes)

	1961–62	*1971–72*	*1981–82*	*1991–92*
Output				
N	418	1,235	1,624	1,807
P_2O_5	224	275	328	...
Total	642	1,510	1,952	...
Consumption				
N	224	405	483	400
P_2O_5	112	110	82	74
K_2O	138	135	112	94
Total	474	650	677	568
Imports*				
N	–171	–885	–1,097	–1,391
P_2O_5	–97	–159	–257	...
K_2O	143	144	92	...
Total	–125	–900	–1,262	...

... Figure not available.
* Negative figures indicate net exports.
Source: Central Bureau of Statistics, *Landbouwcijfers* (Agricultural Figures), several years.

The Netherlands presents another extreme, with a consumption as high as 292 kg per hectare in 1970–71 and 345 kg per hectare in 1980. But thereafter, there has been a tendency to decline, the figure for 1990 being 293 kg. (The consumption figures for India are as taken from respective volumes of *Fertiliser Statistics* published by the Fertiliser Association of India. For the Netherlands, figures are from Central Bureau of Statistics.)

The increase in output in India came from modern large size plants using the best available technology, based on better feedstock, natural gas. This was made possible from recently found reserves of natural gas in the Bombay High and South Basin region. As a result, the share of nitrogen capacity based on natural gas increased from 13 per cent in 1980 to 36.3 per cent in 1987, with a corresponding decline in the share of naphtha from 52 per cent to 33 per cent (see *Kothari's Industrial Directory of India*, 1992, pp. 5–11). In the Netherlands, the availability of large reserves of natural gas has also been an important factor contributing to the growth of the nitrogen fertiliser industry.

We are illustrating the problems and status of pollution control in the Indian fertiliser industry on the basis of the case study of Mangalore Chemicals and Fertilisers Ltd (MCF) in Mangalore, Karnataka. The plant was started in July 1966. It has been producing urea, DAP and ammonium bicarbonate. The output of these products in thousand tonnes during 1991–92 was 262, 147 and 5 respectively. The main raw materials used are naphtha, ammonia (imported) and phosphoric acid. During 1991–92, it used 3,019 million litres of water.

The production process causes both water and air pollution. But the industrial unit has been meeting the standards prescribed by KSPCB. The standards prescribed are given in Table 6.7.

The treated waste water is discharged into the sea through a 4.5 km pipeline during the monsoon (rainy season) and into the forest area of the industrial unit during other seasons. The company informed that the College of Fisheries, Mangalore has been monitoring the MCF effluents since 1975 and found no ill effect on the sea or fisheries. The MCF has installed a hydrolyser stripper unit for carrying out the thermal hydrolysis of urea content and stripping of ammonia from the effluent water. It has also installed a Waste Heat Recovery Boiler. These two alone have meant an investment of Rs 62.8 million. New stacks for boilers and reformers, and

Table 6.7
Standards Prescribed by KSPCB for Fertiliser Industry

A. **Water pollution**	
Ammoniacal nitrogen	50 mg/l
Total Kjeldahl nitrogen	100 mg/l
Urea	200 mg/l
B. **Air pollution**	
Suspended particulate matter (SPM) from boilers	150 mg/m^3
SPM from primary reformer	150 mg/m^3
SPM from urea prill tower	50 mg/m^3
SPM from DAP stack	150 mg/m^3
Fluorine from DAP stack	5 mg/m^3

Source: Questionnaire canvassed with MCFL.

captive power plants for running pollution control equipment have meant another investment of Rs 10.4 million. This can be compared with the total gross fixed assets of the company during 1991–92 to the extent of Rs 1,832.1 million. Being an old company, some of these pollution equipments were added on recently while some others like the pipeline and stacks were built into the design stage itself. The company has its own arrangements to regularly monitor pollution—liquid effluents daily, gaseous emissions monthly, and ambient air quality monitoring weekly at four locations. The company has separate staff for implementing and monitoring pollution control—one manager, one senior officer and two technicians.

Since this is the only major fertiliser producing plant in Karnataka, the information on the cost of pollution control of the industry is the same as for the selected unit, and is already presented in Table 4.10 on industry-level data. The cost burden on account of pollution control was found to be marginal. The annualised costs on pollution control as percentage of total annualised costs were only 4.8, and as a percentage of total sales only 1.9. There is a significant recovery of fertilisers which would have been discharged as pollutants. Though the total economic value of these recoveries has not been estimated by the company, the savings generated from the Urea Hydrolyser Stripper alone amounted to Rs 1.5 million per year. The company has incurred a net loss on the whole during all the three years from 1989–90 to 1991–92. This has not come in the way of implementing pollution control, partly because the anti-pollution law must be obeyed and also partly because the economic recoveries from pollution control are significant.

The fact that MCF, the only major fertiliser producing unit in Karnataka, has implemented pollution control standards does not mean that all units in the industry as a whole have done so. Table 6.8 presents the status of pollution control in operation in the fertiliser industry in India. It shows that 16 out of 105 units (ignoring closed units) in the fertiliser industry have neither implemented nor are in the process of implementing pollution control measures, and there is no information about 35 units on this matter at all.

Table 6.8
Status of Pollution Control in Fertiliser Industry in India (1991)

1. Total no. of units	108
2. No. of units with pollution control in operation	49
3. No. of units with pollution control measures in the process of implementation	5
4. Not implemented	16
5. Units closed	3
6. Units for which no information was available	35

Source: Assocham 1991, p. 19.

The *Dutch* fertiliser industry contributes substantially to the emissions of several substances. Table 6.9 lists the emissions for which the fertiliser industry accounts for more than 10 per cent of total industrial emissions. The environmental policy with regard to the Dutch fertiliser industry includes a wide range of instruments.

Direct regulation takes place in the form of permits for air and water pollution. Air pollution is particularly important for the nitrogen fertiliser industry, which emits large amounts of nitrogen compounds like ammonia and nitrogen oxides. The permits are issued by the provinces, but they have to consider certain national guidelines and decrees. In some cases, deviations from the guidelines are possible, but they have to be motivated by the permit issuing authorities. For new installations in the fertiliser industry, the maximum NH_3 emission concentration is 30 mg/m^3; for existing installations it is 200 mg/m^3. The NO_x emissions from nitric acid

Table 6.9
Emissions from Dutch Fertiliser Industry, 1988

	Relative share in industrial emissions (%)	*Relative share in total emissions (%)*
Air		
Nitrate (as N)	57	57
Ammonia	49	1
Methane	14	0
Phosphates (as P)	12	12
Water		
Phosphates	92	59
Cadmium	73	51
Ammonia	71	71
Mercury	64	23
Lead	53	7
Copper	41	5
Nickel	37	15
Zinc	25	5
Arsenic	21	14
Chromium	18	11
Nitrate	16	0

Source: Emission Registration.

production should not exceed a concentration of 220 (new installations) to 600 ppm (existing installations). The standard for dust emissions (TSP) is 10 mg/m^3 in principle. If the process conditions make an optimal application of filters impossible, this standard can be relaxed to 50 mg/m^3 (cf. the figures for Karnataka in Table 6.7).

Water pollution permits are issued by the State Water Board, as all fertiliser industries discharge in state waters. The environmentally most important issue is that of water pollution by contaminated gypsum from the phosphorus fertiliser industry. There are two fertiliser plants which produce their own phosphoric acid. This process leads to large amounts of gypsum as a by-product. The gypsum is contaminated with several heavy metals, of which cadmium (Cd) is the most important from an environmental point of view. The two firms, which emitted in 1985, 12 and 2.4 tonnes of Cd respectively, had to reduce these amounts to 0.6 tonnes in 1994. It seems that these reductions have been realised indeed, mainly by using raw materials with a lower Cd content and by process changes. The permits also contained the provisions that the firms had to carry out research into the possibilities of minimising the emissions of several substances. They should aim at a complete termination of waste gypsum discharges in the year 2000.

In the area of *communicative instruments*, voluntary agreements (covenants) play a major role. In 1988, the Dutch state had concluded agreements with the two producers, in which the aforesaid reduction scheme for Cd emissions has been laid down. This covenant can be seen as a precursor to the permits. The fertiliser industry also participates in the agreement on the implementation of environmental policy in the chemical industry, which was signed in 1993 by the government and industry. In this agreement, the targets are specified for the emission reductions which have to be achieved by 2000 and 2010. In 2000, for a number of substances the emissions will already have to be reduced by 70 per cent compared with 1985. There are no specific targets for the fertiliser industry but, given their large contribution to several types of emissions, it is clear that they will also have to contribute significantly to the reductions.

The role of *economic instruments* is relatively modest. The fertiliser industry only pays a waste water charge for oxygen binding matter and heavy metals. The total amount of charges paid by the fertiliser industry is unknown, but it is probably only a small part of the total environmental expenses.[9] On the other hand, subsidies for the development of cleaner technologies (especially cleaner production process for phosphoric acid) are quite substantial and can be regarded as a kind of compensation for the stringent demands on this industry.

To conclude, one can say that the fertiliser industry in the Netherlands is a large polluter, which faces a challenging environmental task in an economically harsh period. The Dutch government would not like to see the industry move to other countries with less stringent environmental standards. This would only shift the environmental problems in space. Therefore, the authorities want to keep the communication lines open with the industry. The small number of producers is an advantage in this respect: consultations can be conducted, agreements concluded and monitoring performed relatively easily. The industry is prepared to accept the challenge, in exchange for state support to develop the necessary technological innovations.

6.4 EVALUATION AND CONCLUSIONS ON CASE STUDIES

This chapter presented a comparative analysis of pollution control approaches in three different industrial sectors in India and the Netherlands. The textile, cement and fertiliser sectors present an interesting variety of characteristics, both in terms of pollutants emitted, the processes whereby these pollutants are emitted, and in terms of market structure and organisation. It may be recalled that the theoretical analysis in Part I stressed the importance of these characteristics in determining the suitability of particular types of policy instruments.

The case studies illustrate differences and similarities between industrial pollution control in India and the Netherlands. A striking, but not unexpected, difference is the somewhat laxer approach in India, both with respect to standards and enforcement. The difference in standards may be well justified, given the differences in overall industrial pollution in the two countries, the significant differences in living standards and the associated differences in priority given to environmental issues. Of course, only a rigorous cost-benefit analysis could indisputably reveal whether the differences are economically justified; yet, such a cost-benefit analysis is outside the scope of the present research. Lax enforcement, however, may be harder to justify. Chapter 4 clearly illustrated the budget and manpower constraints of the pollution control authorities in India. Given these constraints, control priorities have been given to medium- and large-scale industries, thereby neglecting small-scale industry to a large extent. That this decision, as such understandable, may lead to inefficiencies is clearly illustrated for the Indian textile sector. In this sector, medium- and large-scale mills in the formal sector of the economy increasingly face competition from small-scale units, mainly power looms. Presently, the power looms produce nearly half of India's cloth (*Kothari's Industrial Directory of India*,1992). The small-scale power looms are already in a favourable position vis-à-vis the mills in the formal sector because they are able to evade labour laws, excise duties, and, of course, pollution control measures. In this situation, every effort to increase the environmental performance of the formal sector will undoubtedly be (partly) undone by an increasing competitive advantage of the unregulated sector. It seems, therefore, to be somewhat futile to rigorously attend to fine-tuning pollution control measures in the formal sector, while any measures to curb pollution are still blatantly absent in the informal sector.

Similarities in pollution control policies between the two countries can also be found. Both countries rely heavily on direct regulation. They both apply a mixture of emission and design standards. However, while the regulatory approach is dominant in the Indian case, supplemented rather weakly with economic incentives, the instrument–mix in the Dutch case appeared to be somewhat more diversified. For example, next to direct regulation the textile industry (and to a lesser extent the fertiliser industry) are confronted with a surface water pollution charge. The Indian water cess is both in its design and its effect not quite comparable with this pollution charge. Notable also is the importance Dutch authorities place on communicative instruments (i.e., covenants) to reduce the pollution burden of the fertiliser industry. However, given the fairly recent history of the covenant instrument it is as yet difficult to predict its effectiveness in terms of pollution control.

The case studies showed some of the complexities and dilemmas of pollution control in the real world. It would, nevertheless, be interesting to confront our real-world cases with the analytical framework presented in Part I. In that part it was argued that specific *situation–characteristics* of the pollution problem at hand decide on the suitability of particular types of pollution control instruments. Three important classes of situation–characteristics were discerned:

- character of pollution;
- character of the polluting processes;
- character of the target group and the market.

These situation–characteristics were then subdivided into 14 specific indicators. In Table 2.1 of Part I these indicators were confronted with 11 different policy instruments and it was indicated which particular combinations of instruments and indicators were likely to be less applicable. For example, large variations in control costs across sources tend to make direct regulation based on uniform

Table 6.10
Situation–Characteristic Indicators and Policy Instruments of the Bovenberg-Framework

Indicators	*Policy instruments*
1. Severity of damage	A. Covenant
2. Non-linearity of pollution damage	B. R&D subsidies
3. Time & space dependency	C. Subsidies on alternatives
4. Few substitutes/high costs	D. Emission charges
5. Market imperfections	E. Product/input charges
6. Innovations are likely	F. Tradable permits('grandfathered')
7. Innovations are unlikely	G. Tradable permits (auctioned)
8. Large variations in control costs	H. Liability
9. Complex processes	I. Direct regulation—emission standards
10. Poor measurability of emissions	
11. Little relation between product/ input and emission	J. Direct regulation—design standards
12. Poor accessibility of target group	K. Deposit–refund
13. Little competition	
14. Dynamic sector (easy entrance)	

standards less attractive in terms of efficiency. Table 6.10 reiterates the 14 indicators and 11 policy instruments.

In applying this analytical framework to the case studies of this chapter, we are immediately confronted with the problem that most industries emit more than one pollutant. In fact, we should use this framework for every single pollutant of every industry because the suitability of instruments depends in part on the character of pollution. One industry may well emit various pollutants with very different characteristics. This presents a major problem which is not immediately answered by the framework. Are the policy instruments equally capable of handling more than one pollutant? A standards based permit can easily contain standards for a list of different pollutants. Covenants may also include a package of pollutants. It is less clear that a charge is effective if it is based on a gamut of pollutants. The inclusion of heavy metals into the surface water pollution charge in the Netherlands did not turn out to be very successful. Is it feasible to confront a firm with a series of different charges? The same questions also apply to the instrument of tradable permits. To have to trade in several permits simultaneously may well increase transaction costs disproportionally and be beyond the administrative capabilities of many firms. Further research will be needed to answer these questions properly.

For simplicity's sake we will select one (major) pollutant per industry. The major pollutant for textiles is bio-degradable waste in waste water (measured in BOD or COD); for cement it is clearly dust; while for fertilisers the situation is more complex. The fertiliser industry emits a large number of pollutants, both in air and water. In the Netherlands the major environmental problem is arguably the release of cadmium into the environment as a result of the production of phosphoric acid. The only major fertiliser producing plant in Karnataka does not produce phosphoric acid, so it does not emit cadmium. The most troublesome pollutants here are probably dust and fluorine emissions.

We will now discuss the *situation–characteristic* indicators for the three industries in Karnataka and the Netherlands.

Textile industry and bio-degradable waste water

1. Severity of damage
   Bio-degradable waste water from the textile mills does not usually cause immediate severe damage. Only in certain circumstances there can be calamities (fish kills). These circumstances should probably be avoided at all costs (see the example of the Gwalior Rayon Silk Manufacturing Company in Kumarapatnam where such an accident caused a public outcry). We will assume that the usual damage is more in the form of a gradual deterioration of the aquatic ecosystem, both in Karnataka and the Netherlands.
2. (Non-)linearity
   There are no indications of severe non-linearity of the damages.
3. Time and space dependency
   The environment impact of oxygen-depleting effluents is dependent on the pattern of discharges. Continuous, relatively low emission levels are less harmful than infrequent discharges of high amounts.
4. Substitutes/costs
   The techniques for pollution abatement are available and their costs are not excessive.

5. Market imperfections
Market imperfections do however exist, especially in Karnataka. As indicated earlier, the mills in the formal sector and the power looms in the informal sector are treated quite differently. This presents a major market distortion. Another market distortion may exist in the form of lack of access of capital for the smaller units.

6/7. Likelihood of innovations
Innovations in control technology are unlikely. This kind of technology has been in existence for a large number of years.

8. Variations in control costs
There are however large variations in control costs. The Indian case study found a range of between Re 1 and Rs 13 per 1,000 litre waste water treated. Indicative for variations in Dutch control costs is that both the case study firms did not install waste water treatment plants, while others did. Yet, all firms in a region are faced with the same charge per unit of pollution. From our analysis, it is impossible to say to what extent the differences in control costs are due to differences in scale of production, and to what extent to other factors.

9. Process complexity
Although textile production involves a large number of different processes, the processes themselves are technically not extremely complex.

10. Measurability of emissions
The level of pollution is not difficult to measure technically. However, in Karnataka emissions are not measured at all for a large part of the industry (the small-scale and informal sector). Small sources of pollution in the Netherlands are also not measured continuously. For calculating the pollution charge, a coefficient table is used to estimate emissions on the basis of process characteristics. If a firm objects to the use of a coefficient table, it has the right to monitor its effluent.

11. Relation between input/product and emissions
The relation between input and emissions is in this case dependent on the application of purification technology.

12. Accessibility of target group
    There is a large difference between the accessibility of the target group in Karnataka and the Netherlands. This is due to the large informal sector in Karnataka which is difficult to approach. No such problems exist in the Netherlands.
13. Competition
    Competition is fierce, both in Karnataka and in the Netherlands.
14. Dynamism
    The textile industry in Karnataka is stagnating and loss-making: the dynamics will not be particularly large. However, there is more dynamism (and easy access) in the informal sector. In the Netherlands too, the dynamics of the sector is not very large.

Table 6.11 confronts the 11 policy instruments distinguished in Table 6.10 with the 13 indicators of situation–characteristics, also presented in Table 6.10. The instruments are in the rows of Table 6.11, the indicators form the columns. In this table, only those indicators are depicted which counter-indicate the use of a particular instrument in either the Netherlands or Karnataka or both. Indicators which counter-indicate the use of a particular instrument in the Netherlands are depicted with a capital N, those that do so in Karnataka are depicted with a capital K. For example, in the cell on the fourth row and the first column a capital K is depicted. This means that indicator 5 presents a counter-indication to the use of emission charges in Karnataka. Indicator 5 is related to market imperfections. Market imperfections in Karnataka are rather severe, especially due to the existence of a large informal sector. As we know from our discussion in Part I (Section 2.2), market imperfections form a counter-indication to the use of emission charges. The other cells in Table 6.11 are filled in an analogous way.

As was explained in Part I, counter-indications are not absolute. If an instrument is counter-indicated for one or more criteria, it does not necessarily mean that the instrument is completely unsuitable. It, however, means that the counter-indication should be taken seriously in implementing the instrument and may require some modifications in its design. Likewise, if a policy instrument is *not* counter-indicated, it does not necessarily follow that the instrument can be implemented without any concern. For example,

Table 6.11
Situation–Characteristics and Policy Instruments for Bio-degradable Waste Water Emissions from the Textile Industry

*Instruments**	*Indicators*						
	5	*7*	*8*	*10*	*11*	*12*	*14*
Covenant						K	K
R&D subsidies		K,N				K	K
Subsidies on alternatives							K
Emission charges	K			K			
Product/input charges	K				K,N		
Tradable permits I	K			K			
Tradable permits II	K			K			
Liability	K				K,N		
Direct regulation I			K,N				K
Direct regulation II			K,N				K
Deposit–refund	K				K,N		

* The policy instruments have been described in Part I.

both India and the Netherlands have no experience with tradable emission permits. Unacquaintedness with a particular instrument may well reduce its attractiveness, both for the industry and for the administrator.

For the Netherlands, the instruments of covenants, emission charges and tradable emission permits are not counter-indicated at first. From the case studies we know that emission charges play an important role in the control of bio-degradable water pollution. As discussed earlier, the surface water charge was not intended to be an incentive charge but a user charge to finance collective water purification. Therefore, a trade-off between charges and tradable pollution permits has never been made.

For Karnataka we see from Table 6.11 that, although there are some counter-indications with respect to some indicators, there seems to be a large scope for flexibility in trying out new instruments, such as covenants. The counter-indications for market based instruments in Karnataka primarily concern the presence of the large, unregulated informal sector. This causes market imperfections and bad measurability of the emissions. The major setback of direct regulation for both Karnataka and the Netherlands is the large variety in pollution control costs. The counter-indications only give first impressions, the final conclusions on instrument choice will be dealt with subsequently.

Cement industry

A similar extensive analysis could be carried out for the cement industry as has been done for the textile industry. However, for the sake of the reader we will keep the discussion brief. The most important pollutant from the cement industry is particulates or dust. High concentrations of dust pose an immediate health hazard to the nearby population. Our framework suggests that economic instruments are less suitable as the pollution damage is severe, moreover, location matters for this industry. The advantage of economic instruments is that they save costs by letting low-abatement-cost producers abate more and high-abatement-cost producers abate less. In cases where the location of the emission of pollution matters, this is clearly unacceptable as it would endanger the population near the high-abatement-cost producer (the one which abates less).

The cement industry also uses much energy. For energy-related emissions (CO_2, SO_2, NO_x), the possibilities for the use of economic instruments are much greater. It is essential, however, to bear in mind that it would be sub-optimal to implement energy-saving measures exclusively for the cement industry. For the greatest efficiency, such measures should apply to all energy users.

The study by Amelrooy (1994) indicates that in the case of the Indian cement industry several factors are favourable for the use of economic instruments. The market is competitive, emissions are readily measurable, and a limited number of large, stationary sources is involved. Therefore, the application of economic instruments in this branch of industry deserves serious consideration.

Fertiliser industry

Since there is only one major fertiliser industry in Karnataka, the case for the use of economic instruments is weak at the state level. With only one plant there are no large differences in control costs. These differences could be larger at the all-India level, which would make the use of economic instruments more appropriate.

The present study, however, does not provide information of the fertiliser industry outside Karnataka. Note the interesting observation in the case study that the economic value of the recoveries of fertiliser by way of pollution control is significant. Communicative instruments should be used to communicate this fact to firms which may have not realised this as yet.

The study by Berg (1994) shows that in the Indian fertiliser industry the accessibility of the target group, the measurability of the effluents and the number of sources are favourable for using economic instruments to abate water pollution. Effluent charges also appeared to be administratively feasible and in harmony with stated policy principles. However, a lack of information on the marginal costs of pollution abatement precluded any firm conclusions on the possible efficiency gains to be obtained by introducing a charge system. Moreover, the fertiliser industry is still heavily subsidised, and this creates a situation in which pollution reduction would not be profitable for the industry under a system of effluent charges.

In the Netherlands, the case for economic instruments to abate cadmium pollution is not strong because cadmium pollution is so serious that only Best Available Technology is acceptable. A charge could be levied on top of the obligation to use Best Available Technology if it could possibly speed up innovations in cadmium recovery echnology.

To wind up part II, we can draw three conclusions. First, the analytical framework for instrument choice developed by Bovenberg seems to be useful in the Indian context too. The second conclusion is that there are both similarities and differences in situation–characteristics between Dutch and Indian cases. Though this suggests that we can learn from each other, we should be careful not to simply imitate each others' solutions. Third, there is no single policy instrument that is most suitable in all circumstances.

With regard to the feasibility of economic instruments in particular, the case studies show that their use in specific sectors may be precluded by factors such as administrative problems or lack of a competitive market. If it is found desirable to introduce economic instruments in order to improve the effectiveness and efficiency of pollution control in these sectors, these impairing factors will have to be addressed simultaneously.

The third part of this volume draws final conclusions on the lessons we can learn from the comparison of policy instruments for pollution abatement in India and the Netherlands.

Notes

1. We have made a liberal use of Amelrooy (1994) for the case studies. References to this source are made wherever specific points have been borrowed.
2. Ginning is more an agro-processing industry than a part of textile industry, and is not included here.
3. Data refer to SBI 22.1 (cotton industry), 22.2 (wool) and 22.4 (textile finishing industry).
4. Chopra et al. found that annualised financial cost varied from Re 0.31 to Rs 5.5 in the cotton textile industry (1993, p. 90).
5. In the wet process, the raw materials (usually limestone) are ground with water and fed into the kiln as slurry. In the dry process, the material is dried, dry ground and fed into the kiln pneumatically. In the former, the wastes (SS, TDS and alkalinity) enter the waste water stream while in the latter, they enter the atmosphere (Amelrooy 1994, p. 63).
6. The data in Table 6.1 refer not only to the cement industry but also to the plaster and other mineral building materials industry. For reasons of confidentiality, statistical data are not available for the cement industry alone.
7. Large sources are defined as sources having a mass flow of > 0.5 kg of dust per hour.
8. 'Arable land' includes net cropped area, area under permanent crops, and current or temporary fallows.
9. The chemical industry as a whole (to which the fertiliser industry belongs) pays some Dfl. 45 million per year on water pollution charges, which is about 5 per cent of its total environmental costs.

Part III

CONCLUSIONS: COMPARISON AND POLICY IMPLICATIONS

7

RETROSPECTIVE SUMMARY

7.1 INTRODUCTION

An environmental policy is an outcome of delicate blending of or balance between economic and ecological considerations. While no country—developed or developing—is today in a position to give up economic development, environmental policy seeks to incorporate the sustainability goal into the process of economic development. Thus each country formulates its own environmental policy and selects its policy instruments in the light of its own specific situation or problems faced. In the first part of the book, it was therefore argued that there could be no general statement about the desirability of any single type of instrument of environmental policy. A framework was introduced there to guide the choice of these instruments.

Three main situation–characteristics were distinguished. The first concerns the nature of the environmental problems: the economic context in which the problems arise (particularly the level of development of the economy), the severity and prevalence of environmental damage and the perception regarding benefits from prevention or abatement of damage. The second deals with the character of polluting products and processes and their alternatives: the availability and costs of clean substitutes, the ease of monitoring discharges, the level and variation in abatement costs between sources. The third aspect relates to the character of the market and institutions influencing environment: economy nexus, including the judicial system. The role of NGOs and people's movements in defence of the environment is also relevant in shaping both the formulation and implementation of environmental policy. In spite of differences in situations and policy across countries, there are a lot of common problems, some of them of global nature. There is a certain degree of commonality in the process of industrialisation and the problems created by it across countries. Thus, we have both differences and similarities which need to be identified.

In the second part of the book, the country-specific context of environmental policies in the state of Karnataka in India and the Netherlands were elaborated. Environmental policy instruments being presently used in both countries were described along with a discussion of their historical and institutional context and the problem that these policies were supposed to tackle. An assessment of these instruments was made in terms of effectiveness and in terms of administrative and financial burden. Aspects such as practical feasibility, institutional compatibility and concordance with policy principles were examined. Special attention was paid to the detailed analysis of the implementation of environmental policies in three branches of industry—textiles, cement and fertilisers in Karnataka (India) and the Netherlands. The framework of Part I was used to evaluate the choice of policy instruments in these industries in both countries.

The third part of the book brings together the results of the foregoing descriptions and analyses. It tries to locate the main differences and similarities between the two countries with regard to the environmental problem, environmental policies, enforcement structure, the role of NGOs and local communities, and the choice of policy instruments. Finally, the study attempts to learn from each other's experiences and tries to infer policy implications. It seeks options for 'new' (combinations of) instruments in either Karnataka or the Netherlands.

The remaining part of this chapter summarises the findings from Parts I and II. Chapter 8 discusses the similarities and differences between India (Karnataka) and the Netherlands, while Chapter 9 looks at the policy implications for both countries, taking into account their specific conditions and contexts.

7.2 SUMMARY OF PARTS I AND II

The present study investigates the constraints and strengths of pollution control instruments in India and the Netherlands, and seeks to formulate recommendations for improvements in the efficiency and effectiveness of such instruments in both countries.

The success of any environmental policy depends on changes in the behaviour of producers and consumers. The environmental authorities can try to bring about these changes by means of various instruments. It is useful to distinguish three broad categories of environmental policy instruments: communicative instruments, economic instruments and direct regulation. The question about which instrument or type of instrument can best bring about these changes depends on a number of factors, which can vary across environmental issues and institutional settings. However, especially after the publication of the report 'Our Common Future' by the World Commission for Environment and Development in 1987 (the Brundtland Commission), the role of economic instruments in environmental policy has been brought to the fore. The concept of sustainable development includes 'merging environment and economics in decision making'. The United Nations Conference on Environment and Development of 1992 has confirmed the need for an 'economic' approach to environmental policy making. For professional economists this approach is nothing new. Already as early as 1920, Pigou had advocated the idea of charges to combat pollution efficiently. From 1960 onwards, and especially since 1975, a vast literature on the theoretical merits and demerits of economic instruments has emerged.

Yet, for various reasons, the penetration of economic and other cost effective instruments in environmental policy making has not matched the high expectation of the early economists. Until recently, western market economies have relied almost entirely on direct regulation or CAC types of instruments. Though these instruments have shown successes, notably in the areas of air and water pollution, they have also shown limitations. They have often shifted problems instead of solving them. They are slow to reach environmental targets, especially in a growing economy; are not very well suited to non-point and mobile sources; and are often inefficient and inflexible.

The response to these and other problems is a shift towards a greater degree of self-regulation of various 'target groups' (e.g., consumers, branch organisations, public utilities) and an attempt to integrate environmental policies across environmental compartments, other policy areas, the life cycles of substances and products and across national borders. This new approach could well favour the use of economic instruments such as charges, tradable emission and production rights, deposit–refund systems and liability.

The Brundtland report was also a milestone in the recognition of the relationship between poverty and environmental degradation in developing countries. The rural environment in developing countries is being threatened by deforestation, loss of fertile land, depletion and degradation of water resources, and loss of biodiversity. Overfishing and water pollution have wiped out fish stocks in many rivers, lakes and coastal waters. In urban areas, air, water and soil are being polluted by emissions from households, traffic, industry, power plants and waste dumps. The proximity between industrial and residential areas and the lack of safety measures create permanent risk of disaster for these people.

The response of governments has been mixed, but generally speaking insufficient to turn the tide. It is only fair to say that their means are far more restricted than in western, affluent societies. Furthermore, poor people often have little choice than to overcharge their environment in order to survive. Widespread bureaucracy and corruption do not favour efficient environmental policy implementation, and finally the debt burden of many countries leaves no room for additional public spending.

Nevertheless, the legislative and institutional framework for environmental policy in developing countries has in many cases been developed along the same lines as in western market economies. Starting from CAC on single pollutant issues, a change towards more integration is already visible, for example, in the Indian Environment Protection Act of 1986.

An increasing challenge and, perhaps, opportunity for environmental policy making in developing countries is their shift towards more economic 'openness' and deregulation. In India, the New Economic Policies (NEP) entail, among other things, fiscal reforms, a decrease of subsidies, and price increases of oil, public transports and other products of public enterprises. Exports are being stimulated by tax exemptions for exporters and devaluations of the rupee. The environmental impacts of the NEP and other Structural Adjustment Programmes are heavily debated. It is essential that environmental policies are in place to counter possible negative effects of these economic reforms without hampering the reforms themselves. The challenge is to make economic development and environmental improvement mutually supportive.

In order to judge the performance of policy instruments, performance criteria should be established. Recently, OECD has distinguished five sets of criteria to assess the performance of environmental policy instruments. Primarily, environmental policy instruments should be able to reach their targets: they should be environmentally effective. Furthermore, they should do so at least cost (including the cost of enforcement), in other words, they should be efficient. They should not severely affect the distribution of income and wealth and should be more or less acceptable for the target groups concerned. Finally, they should not have any adverse side effects on other areas of government policy.

Criteria for environmental policy instruments
Environmental effectiveness
Economic efficiency
Equity
Administrative feasibility and costs
Acceptability

No single type of instrument is superior in every situation. Studies have evaluated the potential performance of different (combinations of) policy instruments in specific policy contexts. A comprehensive framework for the evaluation of policy instruments in different policy contexts is provided by Bovenberg et al. They identify a large number of factors which influence the relative success of an instrument in a specific situation. They call these factors situation–characteristics. They distinguish three main classes of situation–characteristics (character of pollution, character of the polluting process, character of the target group and the market), which can be further subdivided, as indicated in the following diagram.

	Situation–characteristics
Pollution	Severity of damage (Non-)linearity between emissions and damage Time & space dependency
Polluting process	Availability of clean substitutes Market imperfections Variation of abatement costs across sources Complexity Monitoring costs Relation between input/product and emissions
Target group and market	Size and level of organisation of target group Market form Dynamics of the market

The potential performance of policy instruments is affected by these situation–characteristics. Bovenberg et al. do not provide a full theory of the relationships between policy instruments and situation–characteristics; nor do they specify which instrument would be best in a specific situation. They are content to give counter-indications of suitability, i.e., they give warnings that a specific instrument may be less suitable in a particular situation.

In the foregoing discussion, we have stated that for various reasons the penetration of economic instruments in environmental policy has been slower than sometimes expected. Environmental policy makers throughout the world, in the vast majority of cases, have opted for CAC types of instruments. To explain why this is so, it is revealing to examine the specific interests of the various stakeholders in the process of environmental policy making.

Policy makers and administrators are more familiar with CAC types of policy than with other types of instruments. They are of the opinion that CAC types of instruments achieve quick and predictable environmental results. Moreover, in comparison to relatively 'anonymous' instruments such as charges and taxes, their personal influence and room to manoeuvre is greater.

Firms too, usually, prefer regulation over economic instruments because tax revenues are not (completely) channelled back to the industry. In addition, direct regulation can create various opportunities for 'rent seeking'. Finally, direct regulation (e.g., standards) may be more susceptible to bargaining and negotiations than charges and taxes.

Environmental pressure groups usually abhor the idea of 'selling the environment to the highest bidder'. However, in recent years many 'green' political parties and organisations have embraced the idea of a radical tax reform, shifting the tax burden from labour to polluting products.

Consumer organisations and trade unions are usually against price-increasing charges and taxes. They are often unaware that inefficient direct regulations may eventually be even worse for consumer welfare and employment.

It can be concluded that there are strong reasons for the relative modest role of economic instruments in environmental policy. However, their role may become more important once:

- marginal abatement costs rise when standards are tightened;
- more experience is gained in implementing economic instruments: and
- monitoring costs fall.

Economic instruments are also in line with the general thrust towards economic liberalisation and can provide welcome additional revenue for needy governments.

India

The urgency of environmental policy in India is greater than in the developed countries, as it has a direct influence on the survival and livelihood issues for many people. On the other hand, given the widespread poverty there is also an urgent need to speed up economic growth to create livelihood opportunities. The position of the South is difficult. When the North was in a comparable stage of development, it was much easier to externalise the costs of development as environmental concern then was minimal.

Environmental legislation in India dates back to the 1850s. The Indian Penal Code Act of 1860 includes fines and imprisonment for 'fouling water of public springs and reservoirs' and for 'making atmosphere noxious to health'. It also forbids negligent conduct with respect to poisonous substances, combustible matter and explosive substances. Until the 1970s several Acts have been passed containing pollution control provisions. However, there was no institutional set-up to implement and monitor environmental regulation and the approach was rather piecemeal and sporadic.

Some milestones in Indian environmental policy

Year	*Event*	*Aims/Results/Comments*
1853	Shore Nuisance (Bombay and Kolaba) Act	First environmental Act
1860	Indian Penal Code Act	More comprehensive legislation containing several provisions for pollution control
1972	National Committee on Environmental Planning and Coordination	Reviewing of policies and programmes for the environment; environment appraisal of development projects
1974	Water (Prevention and Control of Pollution) Act	Establishment of Boards for prevention and control of pollution of waters at central and state levels
1976	Article 48A of the Directive Principles of State Policies	'The State shall endeavour to protect and improve the environment and safeguard the forests and wildlife of the country'
1980	Department of Environment	Established
1981	Air (Prevention and Control of Pollution) Act	Water Pollution Control Boards were transformed into Central/State Pollution Control Boards
1985	Ministry of Environment and Forests	Established; more powerful successor of Department of Environment
1986	The Environment (Protection) Act	'Umbrella' Act covering water, air, land and the inter-relationships that exist among air, water, land and human beings and other living creatures

Table (contd)

Some milestones in Indian environmental policy

Year	*Event*	*Aims/Results/Comments*
1992	Policy Statement on Abatement of Pollution	Emphasises integration of environmental considerations into decision making at all levels. Includes the principles of Prevention at Source and the Polluter Pays Principle

A breakthrough occurred in the wake of the Stockholm Conference in 1972. The preparatory committee for the conference became the precursor of the National Committee on Environmental Planning and Coordination. In the 1970s, several important pollution control Acts were passed, establishing among other things, the current central and state pollution control boards. In 1980, the Department of Environment was established in the central government, later transformed into the more powerful Ministry of Environment and Forests (MEF). The Environment (Protection) Act of 1986 integrated several pieces of environmental legislation and established the MEF as the apex nodal agency to deal with environmental problems. The 1992 Policy Statement on Abatement of Pollution confirmed the Government of India's commitment to environmental protection and improvement, introducing, *inter alia*, the principles of Prevention at Source and the PPP.

Figure 7.1 shows the institutional set-up of environmental policy in India. The apex policy making body, MEF, acts through the central and state pollution control boards. The Central Pollution Control Board (CPCB) acts like State Boards for Union Territories (which are not under state control, but under central or Union control). It also controls the various State Pollution Control Boards (SPCBs), including the Karnataka State Pollution Control Board (KSPCB).

The main type of instrument of environmental policy is direct regulation. New industries are required to obtain a No Objection Certificate from the relevant SPCB. Before starting the actual industrial process they require Consent for a limited period (a

Figure 7.1: Institutional Structure of Pollution Control in India

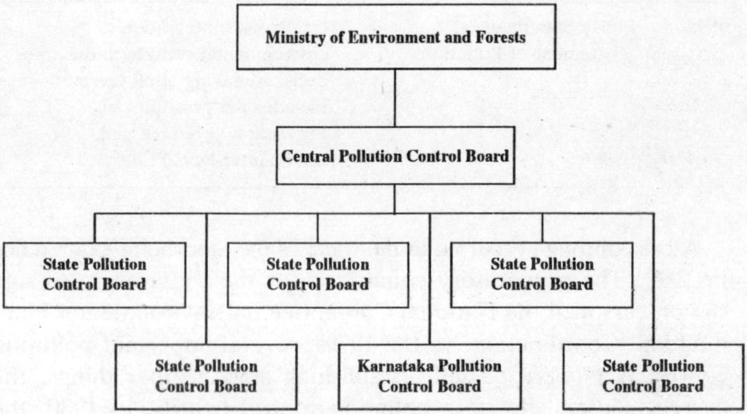

year). Consent can be withdrawn in the case of inadequate compliance with pollution control standards. Industry-specific pollution standards are laid down centrally in the form of MINAS. SPCBs are not allowed to relax MINAS, but they may make them more stringent depending on local conditions.

The frequency of inspection of industrial units depends on their level of severity of pollution. Industries are categorised as 'Red', 'Orange', and 'Green' according to the descending order of the severity of pollution. Before 1987–88, the Boards had little scope for direct administrative action when non-compliance was established. They had to approach a magistrate to enjoin the polluter. Amendments to the Air and Water Acts empowered the Boards to take direct administrative action: i.e., to close down polluting facilities and to stop their supply of electricity and water.

An important factor in envrionmental compliance in India is the right of citizens to take polluters to court. Since the emergence of PIL, whereby citizens or organisations not directly affected by pollution can launch litigation in the public interest, the role of the higher judiciary (the High Courts and the Supreme Court) has become more prominent in pollution control.

Economic instruments do not have a prominent place in Indian environmental policy. Complementary to direct regulation and judicial processes there are some economic incentives to induce firms to implement pollution control measures, including depreciation allowances, tax exemptions and soft loans. There is also a water cess, a levy based on quantity of water consumed by industrial enterprises. The rates of the cess are reflective of the environmental performance of the water user, but they do not give much incentives to abate pollution as their incidence is low compared to the costs of pollution control.

Finally, there are proposals for compulsory environmental audits for industrial units and for the compulsory creation of the function of Chief Environmental Officer in medium and large industrial enterprises.

As to the actual effectiveness of Indian environmental policy there are positive and negative developments. On the positive side there is a clear tendency towards greater integration and coordination, especially since the establishment of the MEF. The increased possibilities of ecological movements, environmentalists and NGOs to take legal action have helped to increase the alertness of PCBs and made it more difficult for polluting industries to ignore environmental regulations. There is also an increasing awareness both on the part of industry and government to shift from end-of-pipe treatment or clean-up technologies to clean technologies, although there is still a long way to go in this area.

Although many industrial units have set up treatment plants, and their share is increasing, the plants may not be operated continuously throughout the year. This may be simply to save on energy costs, or may be due to shortfalls in the supply of energy. For this reason many industrial units still do not comply with pollution standards. In Karnataka in the period 1992–93, among the 99 industries in the highly polluting category, only 38 per cent complied with the standards.

This fact is aggravated by the fact that monitoring by the PCBs is far from complete, due to serious inadequacies of budgets and staff of the Boards. The numerous small-scale industrial plants in the 'Green' category are hardly monitored at all. One of the most glaring lacunae is in checking vehicle emissions, which seriously threaten the air quality of urban areas.

The cost of pollution control measures does not seem to be an impediment to their implementation in the bulk of the cases. As evidence from Karnataka reveals, enforcement costs appear to be only marginal in relation to the turnover and investment of firms. The pollution control expenditure, expressed in annualised cost burden is a small proportion of total cost up to 5 per cent, but mostly within 2 per cent. The distillery industry is an exception, with costs of pollution control expenditure between 10 and 23 per cent. In general, the costs of enforcement and pollution control appear to be easily absorbable and will not make much difference to total costs and profitability.

The prospects of implementation of anti-pollution measures are not quite easy to see due to conflicting influences in operation. On the one hand, environmental laws have become more effective in the 1980s and the overall political mood in the country at higher levels is in favour of a clean environment. The rise of popular movements in favour of the environment is a positive sign, as is the emergence of an indigenous industry to manufacture pollution control equipment. Furthermore, the costs of enforcement and compliance are within absorbable limits in most if not all industries.

On the other hand, there is a disturbing trend of a relatively higher growth of polluting industries vis-à-vis less polluting industries. In the period 1980–81 and 1988–89 the number of industrial units of the highly polluting category increased by 31.6 and 34.4 per cent in India and Karnataka, respectively. The increase in other industries over that period was only 5.7 and 3.2 per cent, respectively.

The possible impacts of the economic reforms introduced in India since July 1991 are difficult to predict. It is feared in certain quarters that in the wake of economic deregulation and liberalisation, pollution control laws will also be relaxed. Some statements of politicians have only worsened these fears. Decentralisation with greater powers to the states, could also have detrimental effects to the environment if the state governments start acting in a less than responsible manner.

A crucial requirement for effective environmental policy is that PCBs can raise a larger proportion of their own finances. This could be done through effluent and pollution charges. The advantage of using proceeds of charges as the major source of finance for PCBs is that they provide an incentive for increased enforcement of anti-pollution law.

The Netherlands

Environmental policy in the Netherlands dates back to the nineteenth century. However, the environment did not become a major public issue until the late 1960s. The first regulatory approach was compartmental and source-oriented, based upon permits. When the limitations of this approach became visible, a shift was made towards the combination of source- and effect-oriented policies (the dual approach) and the integration of environmental policies. Currently, targets and tools of environmental policy are set out in four-yearly Environmental Policy Plans.

Some milestones in Dutch environmental policy

Year	*Event*	*Aims/Results/Comments*
1875	Nuisance Act	To protect man against danger, damage or nuisance from industrial activities
1970	Surface Water Pollution Act and the Air Pollution Act	The first generation of environmental Acts
1971	General Directorate of Environmental Hygiene	Established within the Ministry of Public Health
1972	The Emergency Memorandum on Environmental Hygiene	Important policy document containing a compartmental, source-oriented approach based on permits
1980	General Environmental Provisions Act	First attempt at legislative integration
1982	General Directorate of Environmental Management	Successor of General Directorate of Public Hygiene, now within the Ministry of Housing, Spatial Planning and Environment
1982	Integrated Environmental Programmes	Introducing the dual approach: combining effect-oriented and source-oriented policies
1989	National Environmental Policy Plan	Important policy document aiming at achieving sustainable

Table (contd)

Some milestones in Dutch environmental policy

Year	*Event*	*Aims/Results/Comments*
		development within a single generation's time. Integration of environmental policies
1993	Environmental Protection Act	Completion of legislative integration
1993	Second National Environmental Policy Plan	Shifting attention from goal-setting to implementation problems

The Dutch political system is often described as a 'consensus democracy'. All important policy decisions are usually preceded by intensive consultations with all parties that may be directly or indirectly affected. Hence the strong preference for voluntary agreements and the multitude of advisory bodies. Responsibilities for environmental policy are at the national level, the provincial levels, and the municipal levels. The age-old water boards also play an important role. Finally, in recent times the influence of the European Union in environmental matters has been enlarged.

The main type of instrument in Dutch environmental policy is direct regulation, and especially the permit system. Although there were successes with this approach, especially in abating industrial pollution to air and water, it turned out less effective than expected. Because of this, the 1980s witnessed a reorientation towards a more prominent role for individual responsibility. 'Social internalisation' of environmental responsibility was enacted through the 'target group approach', in which the environmental authorities tried to stimulate environmental responsibility of industry by intensive consultation and negotiation with organised parts of trade and industry. Action-oriented results of these negotiations are laid down in 'covenants' (voluntary agreements) between government and industry.

Environmental policy instruments in the Netherlands

Type of instrument	*Examples*
Direct regulation	Pollution permits, bans, standards for processes and products, compulsory Environmental Impact Assessment
Communicative instruments	Education and information, covenants, environmental management systems within firms
Economic instruments	Charges and taxes, e.g., water pollution charge, charge on surplus manure, fuel charge; tax differentiation, subsidies

Economic instruments play but a small role in Dutch environmental policy. In the early days of environmental policy it was even officially stated that economic instruments should only be applied for less serious environmental degradation. However, user and product charges, complementary to direct regulation, are well established and finance 37 per cent of the costs of environmental policy. The single most important emission charge is the water pollution charge. This charge is levied by the water boards. It is proportional to the pollution load of waste water, measured as bio-degradable matter, suspended solids, toxic substances and heavy metals. Large firms are monitored, small firms and households pay a fixed rate. The charge was set up to be a user charge to finance public treatment plants. However, in practice the charge had an incentive effect, persuading many firms to install private purification plants in order to avoid paying the charge.

Although the role of economic instruments is modest in the Netherlands, wherever used, they have been quite effective in contributing to emission reductions.

Since the National Environmental Policy Plan of 1989, the objectives of Dutch environmental policy are classified into eight policy themes. The following table lists these themes and briefly describes their objectives.

Themes of environmental policy	*Objective*
Climate change	5 per cent reduction of greenhouse gases in 2000 compared to 1991 Phase out CFCs by 1996
Acidification	Deposition target in terms of acidification equivalents (Aeq) is: 2,400 Aeq per ha per year in 2000; 1400 Aeq/ha/yr for forest areas
Eutrophication	Equilibrium manuring in farming by the year 2000; in the period 1985–2000 emission reductions of nitrogen and phosphorous to surface waters of 70 and 75 per cent, respectively
Diffusion of hazardous substances	Objectives generally based on the as-low-as-reasonably-achievable principle
Noise and stench	No growth in number of people affected by noise nuisance until 2000; by 2010 severe nuisance should be negligible; percentage of people affected by stench down to 12 per cent in 2000
Waste disposal	In the year 2000 prevention: 10 per cent of extrapolated trend; recycling: 66 per cent of waste supply
Dehydration	In 2000 a reduction of 25 per cent of the dehydrated area in comparison to 1985
Squandering	No national objectives specified

The total costs of environmental control measures amounted to almost Dfl. 10 billion in 1991. They are expected to rise in the future. NEPP-2 foresees an increase in costs from Dfl. 17.5 billion in 1995 to Dfl. 22.8 billion in 2000 and Dfl.26.5 billion in 2010. The total costs of enforcement in the Netherlands are not known. This study has estimated the annual enforcement costs to be in the order of magnitude of Dfl. 100 million.

Compliance costs of industry average slightly above 2 per cent of net value added or 10 per cent of profits (before taxes). Above average compliance costs befall the oil, chemical, basic metal industries and the public utilities. Except for the leather and shoes industries, profits in other sectors were not substantially affected by environmental compliance costs. Environmental compliance costs are expected to rise in future.

Environmental policy has become firmly established among the main areas of government interference in the Netherlands. The concept of sustainable development has been operationalised through the formulation of mid-term policy objectives for several environmental themes and for 'target groups'. The target groups have a reasonably clear idea of what is expected of them. Direct regulation has been and will continue to be the main category of environmental policy instruments. However, a shift in emphasis is taking place towards enhancing 'self-regulating forces' which can be promoted by the use of communicative and economic instruments in addition to direct regulation.

Case studies

To better compare the working of pollution control instruments in India and the Netherlands, three case studies have been carried out. These case studies compare the working of pollution control instruments in the textile, cement and fertiliser industries in Karnataka and the Netherlands.

Table 7.1 summarises information from the case studies. Some of the main conclusions from the case studies are:

- Both countries rely heavily on direct regulation. However, the instrument-mix in the Netherlands is somewhat more diversified.
- Standards in the Netherlands are stricter than in India.
- The enforcement of standards is sometimes lax in India, especially for small-scale industrial units.
- Compliance costs in both countries are moderate.
- In many cases there is scope to try out new, market based, policy instruments.

Table 7.1
Summary of Case Studies of the Textile, Cement and Fertiliser Industries in the Netherlands and Karnataka

	Textiles		*Cement*		*Fertiliser*	
	NL	*Karnataka*	*NL*	*Karnataka*	*NL*	*Karnataka*
Percentage of total industrial output	0.7	4.2	0.4	3.6	0.9	2.0
No. of units	80	278	29	32	34	9
No. of units in case study	2	7	1	3	1	1
Pollution	Mostly water pollution; almost 30 substances, organic and non-organic		CO_2; emissions to air of particulates and dust		Very many pollutants, both to air and water, including cadmium	
Instruments applied	Permits, effluent charges	National standards	Flexible directives	National standards	Permits, covenants, effluent charge, subsidies	National standards
Level of compliance	Good	None for small-scale units; moderate for larger units	Good	Good; difficulties due to irregular power supply	Good	Good for case study unit; not so good for India total

Cost of compliance (% of sales)	1.5–1.8 (case studies); 0.6 (general)	0.23	0.9	1.6	1.8	1.9
Scope for new instruments	Little scope	Large scope for flexibility in trying out new instruments	Taxes for energy-related emissions	Scope for economic instruments	Not for cadmium	No scope for economic instruments at state level

Source: This study.

8

SIMILARITIES AND DIFFERENCES

This chapter begins with the presentation of similarities and differences between Karnataka (India) and the Netherlands with respect to the nature of the environmental problem. As a policy statement by the Government of India clarifies, the environmental problems in India could be classified into two broad categories: 'a. those arising as negative effects of the very process of development; and b. those arising from conditions of poverty and underdevelopment' (Government of India, June 1992, p. 2). Around 1972 during the time of the Stockholm Conference on Environment and Development, the general perception was that developing countries like India suffer more from environmental problems resulting from poverty and hunger than from those arising out of the side effects of development processes. This was meant to contrast the nature of the environmental problems in developing countries from those in the developed countries, the latter more prominently marked by negative effects of development. With the acceleration of development and industrialisation process in developing countries, countries like India are now being increasingly characterised by negative effects of development, thus reducing the stark contrast between the two types of countries. Yet, the main environmental problems in India are somewhat different in nature. Both poverty and hunger on the one hand and development process on the other have resulted in deforestation; disturbance to wetlands, marine and similar fragile ecosystems; overgrazing; degradation of land; siltation of reservoirs; pollution of air, water and even soils by industries; and generation of toxic and hazardous wastes. Our concern in this study is mainly with industrial pollution.

In the Netherlands, the main causes of environmental problems include: intensive agriculture; excess supply of manure; a high level of industrialisation; large amounts of short-lived products being produced, transported, consumed and disposed of; and a high level of energy use. In the absence of proper abatement measures, these causes lead to degradation of soil, contamination of ground and surface waters, emission of gases, contributing to the greenhouse effect and depletion of ozone layer, and large amounts of acid deposition.

Though India has also been facing industrial pollution to a significant extent, there is a substantial difference between it and developed countries in the level of pollution generated. Despite the fact that India's population is over four times larger than that of entire Europe and two and a half times than that of the United States, the total emissions from fossil fuels and cement manufacture in India were only 178 million tonnes of carbon compared to 391 million tonnes in Europe and 1,329 million tonnes in the United States in 1989 (World Bank 1992, p. 204). In the Netherlands 42 million tonnes of carbon were emitted in 1991 (CBS 1994c). On a per capita basis the differences are more notable. CO_2 emissions (in terms of carbon) were only 0.21 tonne per capita in India in 1989, but 2 tonnes in Europe, 2.80 tonnes in the Netherlands and 5.34 tonnes in the United States. In terms of CO_2 emissions per million dollars of GDP, India fared worse. This figure was 670 tonnes for India, 177 tonnes for the Netherlands, and 259 tonnes for the United States (calculations based upon CBS and World Bank). The magnitude of the pollution problem being what it is, it is understandable if India has a little laxer attitude towards pollution abatement than Europe, though admittedly complacency in the matter is not desirable. However, while less strict standards of pollution abatement in India could be considered reasonable, their lax implementation cannot be considered excusable. More of this later.

Though environmental concern can be traced back to earlier years in both countries, it is interesting that both took major initiatives in the early 1970s. In both countries, major pollution abatement measures started with water, extending later to air. In both countries, the earlier piecemeal measures gave rise to a more integrated policy culminating in Environment Protection Act (1986) in India and Environment Protection Act (1993) in the

Netherlands. It is noteworthy that India has taken these steps in time in spite of much lower levels of income and development.

In both countries, there is a strong environmental lobby and NGOs are very active in taking up environmental causes and in leading popular movements. This has made the government and implementing agencies more alert to the needs of environment and has also served as a countervailing power against private industries which may prefer a lax policy. In India, NGOs have also taken up the constructive role of actual eco-development, and have been playing an active role in reforestation and restoration of grazing lands. Indian NGOs and concerned citizens often pursue their environmental objectives by judicial means (PIL) whereas the Dutch NGOs are more involved in the formal policy making process (they are for instance represented in advisory councils for the government).

The environment policy principles and objectives in the two countries read alike (see Section 4.1 on India and Section 5.1.4 on the Netherlands). Both countries have adopted in their constitutions the protection of the environment as a governmental task. Both have accepted the PPP, fixing the responsibility for pollution and pollution control on the polluter. Both accept the application of best available and practicable technology. In the Netherlands, the policy is more clear on this, spelling out 'reducing the most polluting substances with the best technical means (regardless of costs), and less harmful substances with best practicable means (i.e., taking into account 'reasonable' costs)'. Though the exact wording of policy statements may differ, both countries aim at pollution abatement at source as well as in the environment (effect-oriented). India has standards laid down for pollution at source and also ambient standards, and has taken special pains to reduce pollution in heavily polluted rivers like the Ganga. The Netherlands has clear targets for reducing pollutants by the stated years as laid down in the National Environmental Policy Plans. The first such plan was published in 1989. Though the environment figures in the Five Year Plans in India along with other sectors, physical targeting is much less clear for the environment in India than for other sectors, and also compared with the pollution abatement targets in the Netherlands for ambient air and water quality. Apart from several river stretches, urban air quality has also deteriorated in many Indian cities drastically, requiring clear targets for action. It is noteworthy in this respect that in 1995 India introduced lead-free petrol and has insisted on catalytic converters in newly produced automobiles to reduce air pollution in cities.

There are similarities and differences in the enforcement structure too. Both countries rely more on direct regulation than on economic incentives, though both these instruments are used in the two countries. However, economic incentives seem to be used more effectively in the Netherlands than in India. But first let us discuss regulation. The earlier approach in India was judicial but since the mid-1980s, the shift is clearly towards giving more initiative, scope and power to administrative agencies in charge of regulation, cutting down delays and encouraging stricter enforcement. This has brought India closer to the Netherlands. In India, however, there are Minimum National Standards which are uniform, but the SPCBs have the powers to make the standards stricter in their respective states depending on the situation. The standards are uniform across the industrial units within a state. In the Netherlands on the other hand, the pollution control authorities can issue individually tailored permits which can enable stricter standards where either the pollution effect is more worrisome or the concerned industrial unit is more efficient and can afford to implement stricter standards. Covenants between the authorities and industry to achieve targeted pollution abatement are also common in the Netherlands, where more ambitious emission reductions are tried to be achieved than is possible under a purely regulative system.

Such fine tuning is not evident in India. In fact, the Indian system does not encourage much discretion to authorities due to the fear of corrupt practices. Even though the PCBs have now been given the administrative powers to order the closing down of non-compliant industrial units and/or to stop the supply of electricity and water to them, the units can go in appeal against the PCBs to judiciary. Fortunately, however, the judiciary, particularly the apex courts (Supreme Court at the national and High Courts at the state levels) are quite sympathetic to the cause of the environment and have sometimes taken initiatives to order EIA reports and take other necessary actions. The role of the judiciary has become more prominent than before, even with the increase in the administrative powers of PCBs, particularly with the introduction of PIL and the right of information granted to citizens. The judiciary and the administrative network have complemented each other to operate the regulatory systems.

In the Netherlands, the administrative and the judiciary system are also complementary in the enforcement of environmental laws and regulations. Administrative enforcement is mainly directed at compliance with the rules whereas criminal sanctions are used against intentional or grossly negligent polluters, recidivists, and in cases of damage to the environment. In the latter case, compensation of the damage can also be claimed on the basis of private law. This possibility is being used by the Dutch government in numerous cases of soil pollution.

In addition to the regulatory system, both India and the Netherlands have been using a few economic or financial incentives which are expected to have a beneficial effect on the environment. Neither country has gone so far as to replace the current system of taxing income and labour by a system of taxing the use of the environment. The measures are rather piecemeal and one of a nature supplementary to the regulatory system. The policy statement for Abatement of Pollution is emphatic on the need for 'increase in the development and application of financial incentives' (Government of India, February 1992). The economic incentives, including taxes and subsidies applied in both countries which have environmental significance, have been enumerated and discussed in the respective country studies in Part II. To recall briefly, water cess, tax rebates and concessions, and soft loans for installing pollution control equipment are the main economic instruments used in India. In the Netherlands, there is a more comprehensive variety—waste water charge, charges on surplus manure, aircraft noise charge, general fuel tax, waste and cleansing charges, tax differentiation and varius subsidy schemes. The waste water charge, which was imposed mainly to meet the expenses of Water Boards, has been instrumental in reducing the amount of pollution discharged into water. On the other hand, the water cess in India has a nominal incidence and has no such environmental impact at present. Though the rates vary between firms complying with the Water Pollution Control Act and those not complying, the water cess meets only the limited purpose of meeting a part of the expenses incurred by the PCBs.

Do the costs of enforcement and compliance impose too much of a burden on the respective economies of the two countries? This question is particularly relevant in the case of India having a low level of income and development. Fortunately, however, the burden seems bearable in both the countries. Let us first take the enforcement costs incurred by the respective governments. For the Netherlands, the annual enforcement costs of environment policy were estimated to be in the order of Dfl. 100 million or US $53 million. This amounted to 0.023 per cent of the net national product of the Netherlands. Estimating these costs for India at the national level is difficult as figures could not be collected for the constituent states, and also at the national level. The enforcement costs of industrial pollution control incurred by the KSPCB were Rs 57.2 million or US $2.3 million in 1991–92. These figures are on the lower side, because they do not include the expenses incurred by the CPCB in Karnataka, and also expenses incurred by the state on other environmental sectors like arresting deforestation. However, in the absence of more comprehensive figures, this figure shows that the enforcement costs amounted to 0.027 per cent of the state domestic product. This proportion is close to the similar proportion in the Netherlands, but could be higher if the estimates of expenses incurred in the state by the CPCB are included. This proportion could be considered well within the bearable range. It should be appreciated, however, that in spite of a lower level of income and development, Karnataka spends a slightly higher proportion of its state domestic product than the Netherlands on the enforcement of environmental regulations.

Compliance costs incurred by the firms vary depending upon the industry concerned, and even within a given industry between firms depending upon the age of the firm, technology adopted and so on.

The annualised costs of pollution control varied in Karnataka from 0.2 to 1.9 per cent of turnover among the selected firms, and from 0.2 to 2.36 per cent in the Netherlands. It is remarkable that the range levels are close to each other in the two countries. We consider this to be well within the affordable limits of the concerned firms. However, the actual compliance costs may be higher in Karnartaka than what is indicated above, considering the somewhat laxer implementation of the standards. The enforced standards are stricter in the Netherlands, and yet the firms in the country have as low compliance costs as in India relative to turnover. This suggests that the pollution treatment technology in the Netherlands is more cost effective.

9

OPPORTUNITIES FOR IMPROVEMENT IN POLLUTION CONTROL POLICIES

9.1 SHORTFALLS AND POLICY IMPLICATIONS: INDIA

Though the implementation of pollution control in India has been much more effective during the 1990s than in the 1970s, there are still shortfalls and constraints affecting it. A much greater proportion of firms have set up pollution treatment plants for both air and water pollution now than before, and the new firms have been trying to build it into the production process or design itself so that pollution control is more automatic and cost effective. Yet, old firms with add-on technology still dominate the scene and have not been very effective in pollution control. The problem partly lies with monitoring and the enforcement of the standards. Many firms install the treatment plants but do not operate them fully during the entire period of production, and untreated effluents are released into the environment. According to the Annual Report of the KSPCB for 1992–93, 62 per cent of industrial units in the highly polluting category did not comply with the standards. Of the 30 parameters of pollution in the textile industry, only five are monitored. In this case, it is hard to hope that the firms try to maintain standards where they are not even monitored. Even where standards are monitored, the powers of the Board are sparingly used. In spite of a much larger number of non-compliant firms, the KSPCB ordered the close down of only five firms under the Water Act and four firms under the Air Act.

Most of the standards are based on concentration of pollutants in effluents and emissions. These can be met by diluting the concentration, which does not reduce the mass of pollutants and increases the wastage of water. A shift to mass based standards is mooted now, but it may take some time to be implemented, and needs closer monitoring.

Unfortunately, however, the PCBs are chronically short of funds for effective functioning. Only Rs 289 (about US $9.3) were available on an average to monitor one industrial unit in 1993. The PCBs rely largely on funds generated by them through the imposition of the water cess, fees and fines. The grants from the central government contributed only about 13 per cent of the receipts of the KSPCB in 1992–93. With the pressure to keep down budget deficits under the new economic policy, the SPCBs cannot hope to get much from the central government and will have to rely mainly on own funds collected from the industries. The water cess being levied under the Central Act has to be first credited to the CPCB, which is subsequently followed by reimbursement. This involves avoidable delays and a way may have to be found whereby the SPCBs can keep the proceeds of the water cess directly for themselves.

With the resources of PCBs being so limited, it is no wonder that they concentrate more on large- and medium-scale industries. The small-scale industries are too numerous to be monitored properly. It is also uneconomical for most of them to operate their own pollution treatment plants. The PCBs have, however, been promoting common or collective treatment plants for treating pollution generated by the small-scale units. Where they are absent, they release effluents into drains, leaving it to the municipalities to treat them. Since the municipalities are also short of funds, they should recover the expenses of such treatment from the polluting firms.

The shortcoming of PCBs is to some extent compensated by the alertness of the public or NGOs who have launched PIL in several cases. They have not spared non-compliant industries, whether big, medium or small scale. Even small-scale tannery units which were highly polluting had to be closed down in Kanpur in northern India following PIL and the Supreme Court Order. Such things have taken place elsewhere too. However, the provision for PIL can only serve as a deterrent to polluting firms and cannot take the role of enforcement agencies like the PCBs. It also takes a long time to get redressel from pollution under PIL. The possibility of PIL can make the PCBs more alert, but unless the aforesaid constraints of PCBs are removed, they cannot be fully effective.

To begin with, the PCBs may have to shed some unnecessary advisory burdens. They somehow assume the role of indicating what technology to adopt to reduce pollution up to the standards prescribed, if the industries are unable to cope with pollutants to the extent expected. Though in principle, the best available technology has to be adopted, it has also to be economically viable, i.e., it should not be unreasonably costly for the firm. If the PCB is unable to indicate or advise on such a pollution treatment technology, it feels obliged to be indifferent to the non-compliance of the firms, and the firms concerned also feel morally and legally justified in continuing non-compliance.

This problem as well as the problem of inadequate funds available to PCBs in India could be met by imposing a deterrent charge on effluents and emissions that are still untreated and contain pollutants. A deterrent charge of this nature would force the firms to seek a proper technology from consultants in the field and adopt it. For this, the charge would have to be such that it is cheaper to comply with the standards prescribed than to pay the charges. To meet this criterion, the charge would have to be equal to the marginal cost of pollution up to the standards prescribed. The advantage of such a charge is that it introduces a certain amount of automaticity in pollution abatement, somewhat reducing the burden on the regulating agencies in the bargain, and also achieves pollution abatement at less cost. A reliance on regulation alone and on the accompanying judicial process involves long delays. The charge is proposed not in replacement of the regulatory framework of enforcement but as a supplement to it as in the Netherlands. The charge has to be on all the pollutants released untreated even after the standards are met, if there has to be a continuous incentive for firms to adopt a more cost effective technology which can reduce pollution beyond the standards prescribed. A more efficient firm can achieve pollution abatement even beyond the standards prescribed, as it equates the marginal cost of abatement with the pollution charge at a point beyond the standards prescribed. A less efficient firm would reach this point before achieving the standards prescribed, but it must meet at least the standards nevertheless under the regulatory law. This would force the firms to seek more efficient technologies and build pollution prevention into the process or design itself, and can make industrial production environmentally sustainable. It is creditable that the Policy Statement for Abatement of Pollution has already accepted the need for adoption of such economic instruments like the effluent charges (Government of India, February 1992).

Some countries like Germany and Italy have differential charges. In Germany, discharges meeting or exceeding the state-of-the-art effluent standards have to pay only half the normal rate. In Italy, the charge is nine times higher for firms that do not meet the prescribed standards than for firms that meet them (Tietenberg 1990, p. 20). In India, a working principle could be adopted to charge two times the marginal cost of meeting the standards to non-compliant firms and only half the marginal cost to compliant firms on effluents/emissions released untreated.

A complicating factor which makes the imposition of an effluent charge somewhat difficult is that often more than one pollutant is involved. The charge, to be effective, has to based on all the pollutants, the release of which into the environment is undesirable. Thus, it should not be a single charge, but an amalgam of charges on each pollutant or at least each important pollutant. If in abating one pollutant, another is automatically abated, only one of such pollutants could serve as a charge base. Since there is always a technical possibility and also temptation to transform one form of pollution into another, it is important that all forms of pollution are taxed together.[1] For example, solid waste pollution can be burnt and transformed into air pollution. Scrubbing of stack gases can reduce emission of particulates into air, but can increase water pollution. In practice, it would be quite a complex exercise to calculate the marginal cost of abating each pollutant at the standard level and add them up. A judgement based on trial and error may have to go into the fixation of the level of the charge, helping the policy maker to continuously arrive at an optimal charge. This difficulty of multiple pollutants also requires that a charge on pollutants, even an amalgam of pollutants be complemented by direct regulation banning the release of any effluents beyond the standards prescribed, so that both the instruments act in a mutually reinforcing way, plugging the loopholes if any.

Another complication in calculating the required charge is that there are several firms in an industry and several industries in a region. It may be difficult and also economically inefficient to operate a system of charge rates tailored to each industrial unit. The marginal cost of meeting the standards in an average or a representative firm may be a reasonable proposition often, if not in all cases.

Economic instruments are relatively easy to operate in the case of dealing with vehicular emissions. India has already introduced lead-free petrol, but the price difference between unleaded and leaded petrol should be high enough to act as an incentive through a heavy tax on leaded petrol. Similarly vehicles without catalytic convertors could be taxed sufficiently high to induce their fixing.

Some economic incentives are already in operation in India which have been discussed in Section 4.2. The water cess, as has already been stressed, is not operated as an economic instrument of pollution control policy in the nature of the charge on effluents in the Netherlands or in the way described as necessary in the preceding paragraphs. The burden of water cess today is nominal as shown in Table 4.2 in Part II and does not provide an effective incentive as it is cheaper to pay the full cess than to abate pollution. The maximum difference in the rate of water cess between complying and non-complying firms is 4.5 paise per kilolitre (Re 0.045), while the average cost of waste water treatment in the textile industry was about Rs 2 per kilolitre. This can also be compared with the cost of treating waste water incurred by the Bangalore Water Supply and Sewerage Board (BWSSB) which was Re 1.05 in 1989–90. Since the Board did not quite meet the effluent standards of KSPCB, this is an underestimate and the required cost could be much higher. Moreover, the cost figures for both the textile industry and BWSSB are average costs, but the marginal cost—on which the effluent charge has to be actually based—is expected to be much higher than the average cost. All this shows how nominal the water cess is and cannot be considered at its present level as an effective economic instrument. Though the water cess fails the test of environmental effectiveness, it is easily administrable and because of its low incidence has been found acceptable to industries. If a charge on effluents and emissions were to replace the water cess, it may be environmentally effective but may not be easily administrable and acceptable to industries. If it is not administratively feasible, then it may not also be environmentally effective. This is a dilemma which India may have to resolve in the coming years. However, it seems that India is unlikely to move in favour of a large-scale shift over to effluent charges immediately. Instead, it may continue to rely more on regulations, fees, fines and threats of orders for closure of firms. If, however, effluent/emission charges are in the long run more suitable, as it seems theoretically, India has to at least make some experimental efforts in the direction.

Irrespective of whether pollution abatement takes place through regulation or pollution charges, it is clear that in either case it is the polluter who has to pay the cost of pollution abatement. It is highly inequitable to shift the cost of pollution to victims of pollution. The PPP has been accepted both in India and the Netherlands.

Both countries, however, also use positive economic incentives, if not direct subsidies as an instrument of economic policy. Direct subsidies are inconsistent with the PPP. The role of reliefs could be increased particularly when charges on untreated or unabated pollutants are levied. For example, firms shifting over from a less effective add-on technology to a more effective built-in technology of pollution prevention, could be given soft loans on investments involved. At present, soft loans are extremely limited in scope and amount in India, confined only to loans given by the Industrial Finance Corporation of India. The commercial banks have not been extending soft loans for investing in pollution abatement technologies. The bureaucratic hassles involved in claiming subsides and reliefs also have to be reduced.

In any case, there will have to be a stricter enforcement of environmental regulations and standards in India even if the standards are less stringent than in the developed countries. The world has become increasingly watchful now and if we are not strict and efficient in implementing what we ourselves have set as standards, it could be interpreted as amounting to a trade malpractice or at least a hidden subsidy in international trade, inviting sanctions and trade restrictions. This hardly goes with the goal of economic reforms and an increasing integration of the Indian economy with the world economy.

Will this impose too much of a burden on the economy of the nation and the country? In the preceding section we have already noted that this is not so. In fact India has been spending more on pollution control relative to its income than even some of the developed countries, but the technology has to be more efficient and cost effective. The international community of nations would also come to India's help in this regard and meet their obligations.

9.2 SHORTFALLS AND POLICY IMPLICATIONS: THE NETHERLANDS

Looking at the Dutch experience with regard to abating and preventing industrial pollution in the last two decades, we may conclude that significant improvements have been achieved in terms of effluent and emission reductions. Whereas real GDP increased between 1970 and 1990 by over 60 per cent, the industrial discharges of most air and water pollutants have increased at a much lower rate or, in some cases, even decreased substantially. Nevertheless, there is no reason for complacency. Figure 5.3 in Part II clearly shows that there is still a long way to go before reaching something like a 'sustainability level'.

The past achievements have mainly been realised by means of direct regulation. Economic instruments have played an important role only in the case of surface water pollution charge. In recent years, the emphasis has shifted towards voluntary agreements and self-commitments. The fact that industry has developed a more cooperative attitude can be explained by a combination of a sense of responsibility, concern about the corporate image, and fear for more regulations and liability claims.

As has been noted in Part II, a much wider use of economic instruments in Dutch environmental policy does not seem very likely in the short term, although the discussion on 'ecologising' the tax system is still going on. For the time being, the introduction of new environment-oriented charges and taxes will probably be limited to items like waste, energy and raw materials. Tradable permits are still being discussed, although their implementation requires several institutional adaptations. More generally, however, there will be attempts to arrive at more cost effective ways of reaching policy objectives. The 'cheapest' opportunities for pollution reduction have by now been exhausted, and some industries are looking for possibilities to fulfil their obligations in another way than by reducing their own emissions. The Dutch electricity producers, for example, have contributed to the financing of a flue gas desulphurisation installation at a Polish power plant. They have also embarked upon an afforestation programme in various parts of the world, in order to compensate for part of their CO_2 emissions. Such initiatives could be followed in the years to come, for instance by 'joint implementation' projects which are envisaged under the Framework Convention on Climate Change. On a national level, on might also expect exchanges in emission reduction obligations between firms or branches of industry. Rather than in a formal tradable permit system, these exchanges could take place within the framework of voluntary agreements and self-commitments.

Obviously, such exchanges should only be allowed when the exact location of the emission is irrelevant for the environmental impact.

Despite these possibilities for cost minimisation, it seems clear that generally speaking the costs of pollution prevention and abatement, which as we have seen are presently still quite low, will keep rising for most industries, both in absolute and relative terms. The extent of this increase depends mainly on the following factors:

- the increase in the volume of industrial activity;
- the development of technology;
- the environmental objectives aimed at.

This can be illustrated through Figure 9.1. On the horizontal axis, the level of environmental 'cleanliness' is depicted, and on the vertical axis the costs of pollution prevention and abatement. Given the present state of technology and level of production, the marginal costs of emission reduction are given by curve 1. The marginal costs of attaining environmental quality level A are X. In the future, technological development is expected to lead to curve 2 (assuming a constant level of production), allowing a better environmental quality level B at the same marginal cost level X. However, increasing production will shift the curve back to curve 3, so that environmental quality level B can only be attained at marginal cost Y.

Industry in the Netherlands is concerned about environmental costs becoming disproportionally high in comparison with competing countries. The Dutch economy is very open, and cost increases can often not be passed on to the customers on competitive international markets. In designing its environmental policy and choosing the instruments, the Dutch government should take these concerns into account. Studies on the economic consequences of a substantial and comprehensive energy tax indicate that its coordinated introduction in all OECD countries would be preferable to a unilateral action by the Netherlands. At the same time, it would be unwise to aim at a complete international harmonisation of environmental objectives, standards and instruments. There are differences in natural conditions, levels of development, technological possibilities, and preferences and priorities between countries, and these differences may justify differences in environmental

Figure 9.1: Impact of Production Growth and Technological Development on Environmental Costs

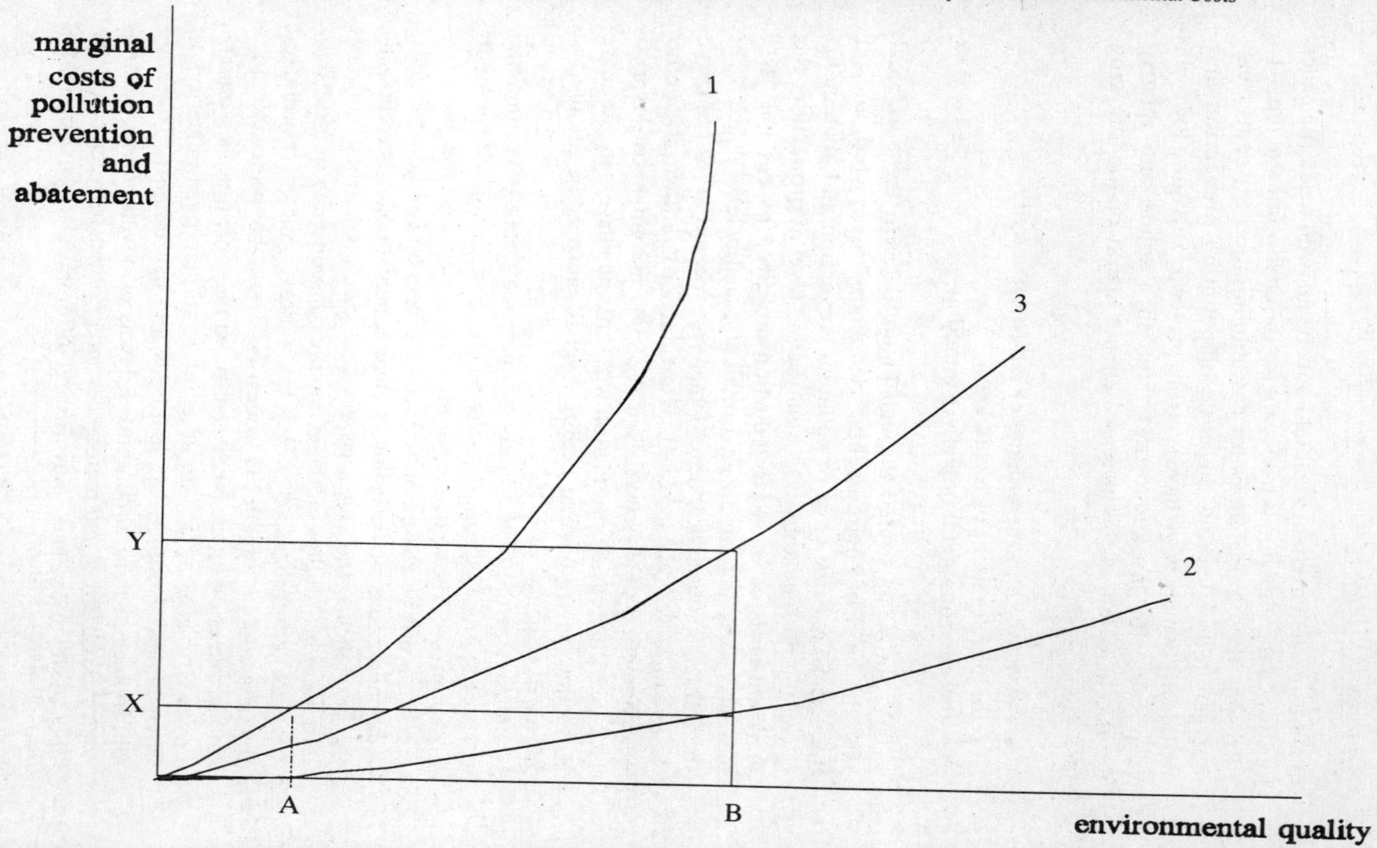

policy. Higher environmental costs can often be compensated for in terms of productivity increases, better product quality, higher service levels, improved corporate image, etc. Furthermore, stringent environmental standards and/or high emission and effluent charges will stimulate the development of new technology, which creates business opportunities for innovating industries, especially if other countries follow the same policy with some delay.

Nevertheless, there will be cases in which relatively high environmental costs (usually together with other factors) lead to the relocation of certain industries in countries with less stringent demands. In principle, this should not be a reason for refraining from the high standards or charges needed to achieve the desired environmental objectives. Instead, the country to which the industry moves should be aware of the possible environmental consequences of attracting polluting industry. A long-term perspective is essential in this respect. The authorities should realise that their citizens will not accept ever increasing levels of pollution, the less so as their standard of living rises. At the same time, industry should take into account the possibility that the perceived cost advantage will be a short-lived one, as environmental demands (including their enforcement) in developing countries will become more stringent in the course of time.

9.3 DIVERGENT ROUTES TOWARDS A COMMON GOAL

Finally, let us return to the two objectives of this book presented at the outset:

1. to provide insight into the constraints and strengths of pollution control instruments in India and the Netherlands; and
2. to recommend improvements in the efficiency and effectiveness of policy instruments aimed at abating and preventing pollution in India and the Netherlands.

To what extent have these (ambitious) objectives been met?

The analysis and assessment of policy instruments in both countries, as presented in Part II of this book, has certainly contributed to the first objective. Despite their large differences, both countries appeared to share a number of strengths and weaknesses in their approaches towards pollution control. There are also several issues which are more characteristic for one of either, as a result of the specific conditions prevailing in that country. Some of the main features can be summarised as follows:

	Strengths	*Constraints/weaknesses*
India	comprehensive set of ambient and emission standards important role for Public Interest Litigation wide range of enforcement instruments available	lack of quantified targets for the long term standards cannot be adapted to specific situation inadequate resources for Pollution Control Boards
The Netherlands	strong planning approach industry involvement in execution of plans some effective economic instruments (e.g., waste water charge)	implementation lags behind effective but costly measures often abandoned for reasons of international competitiveness possibly too much reliance on voluntary agreements
Common	environmental protection constitutional task for government	insufficient attention and tools for reduction of pollution from small firms and non-point sources
	generally accepted policy principles (such as PPP) strong environmental lobbies and NGOs	scope for use of economic instruments not exploited inadequate monitoring and enforcement

When we try to translate these strengths and constraints into policy recommendations (the second objective), it should come as no surprise that these recommendations are not identical for India and the Netherlands. To be sure, there is room for a more intensive use of economic instruments in both countries. Nevertheless, the specific situation–characteristics prevailing will determine the appropriateness of applying economic instruments to deal with a particular pollution problem in a particular industry. Moreover, a number of specific conditions will play a role in the design of the instrument. The revenues of new environmental taxes and charges, for example, could be used in India to reinforce the financial capacity of the PCBs, whereas in the Netherlands it is probably more appropriate to allocate such revenues to the general public budget.

More attention for monitoring and enforcement is needed in both countries. This is not only a matter of sufficient (financial) resources. In India, for example, tasks and responsibilities of the PCBs as well as existing procedures could be redefined in order to enable them to work more efficiently. The focus of enforcement activities should shift more to the small, 'informal' firms. In the Netherlands, the frequency of checks and inspections seems quite low in general, and a better coordination of enforcement activities is needed (see Section 5.4.2).

Thus, although India and the Netherlands share the goal of sustainable development and implement a number of environmental policy principles, their routes to achieve that goal and to implement those principles may differ. Each country has to base its choices in pollution abatement policies on a careful analysis of the specific situation and conditions. However, in doing so, they may still learn from each other's positive and negative experiences.

Note

1. This is true of regulation too. It has recently come to light how thousands of dyeing and dye related industries in Panipat district of Haryana 'solved' the problem of surface water pollution by simply dumping the effluents underground through bores drilled into aquifers. The ground water in parts of the district has been rendered unfit for human consumption (*Down to Earth*, 28 February, 1994, pp. 5–8).

ANNEXURE: A COMPARISON OF ENVIRONMENTAL STANDARDS IN INDIA AND THE NETHERLANDS

A. AMBIENT QUALITY STANDARDS

1. Air (in μ/m^3)

	India				*The Netherlands*			
	Sensitive areas		*Other areas*		*Target value*		*Limit value*	
	mean	98%	mean	98%	50%	98%	50%	98%
SO_2	30		80		30	100	75	250
NO_2	30		80					135
CO	1,000		10,000					6,000
Pb	1.5		1.5				0.5 (mean)	2
SPM	100–300	400–1,000	200–500	400–1,000			30	90

Note: 50- and 98-percentiles indicate that the values stated should not be exceeded in 50 and 98 per cent of the observations, respectively. Numbers of observations and averaging times required differ between standards.

2. Water

	India	*The Netherlands*
pH	$7.0 \leq pH \leq 8.5$	$6.5 \leq pH \leq 9.0$
Suspended solids	–	$\leq$ 50 mg/l
Nitrate	$\leq$ 45 mg/l	$\leq$ 10 mg/l
Phosphate	–	$\leq$ 200 μg/l
Cd	$\leq$ 10 μg/l	$\leq$ 1.5 μg/l
Cr	$\leq$ 50 μg/l	$\leq$ 50 μg/l
Mineral oil	–	$\leq$ 200 μg/l
BOD	–	$\leq$ 7 mg/l

Note: Standards apply for surface waters which should be suitable for preparing drinking water.

B. EMISSION AND EFFLUENT STANDARDS FOR THE TEXTILE, CEMENT, AND FERTILISER INDUSTRIES

	India	*The Netherlands*
1. Air		
Cement industry:		
Total dust	250–400 mg/m^3	10–50 mg/m^3
Fertiliser industry:		
Particulate matter	50–150 mg/m^3	10–50 mg/m^3
Fluorides	25 mg/m^3	5 mg/m^3
2. Water		
Fertiliser industry (phosphate fertiisers):		
Phosphates (as P)	5 mg/l	70–600 mg/l
Fluoride (as F)	10 mg/l	70–3,000 mg/l
Total chromium	2.0 mg/l	0.2 mg/l
Textile industry:		
Suspended solids	100 mg/l	20 mg/l
Oil and grease	10 mg/l	20 mg/kg

Note: Effluent standards for the Dutch fertiliser industry have been taken from individual permits. The relatively high values for phosphates and fluoride compared to India may be explained by the fact that these permits also contain limits for the total discharges.

BIBLIOGRAPHY

Abubacker, M.H. (1990). Socio Economic Impact of Pollution: A Study of Cement Factories in Tiruchirapalli District, Tamil Nadu, Ph.D. Thesis submitted to Bharathidasan University, Tiruchirapally.

Adriaanse, A. (1993). *Environmental policy performance indicators*, SDU, The Hague.

Amelrooy, M. Van (1994). *Indian Environmental Policy and the Use of Economic Instruments*, IDPAD OP&R 1994–41, New Delhi.

Arentsen, M.J., J. Th. A. Bressers and P.J. Klok (1993). Van urgentienota naar NMP en verder (From Emergency Memorandum to NEPP and beyond), in Raad voor het Milieubeheer (1993). *Milieu van jaar tot jaar 1992–1993—20 jaar Nederlands milieubeleid* (20 years of Dutch environmental management), pp. 13–49. Delwel, The Hague.

Associated Chamber of Commerce and Industry of India (Assocham) (1988). Environmental Legislation and Industrial Development, Background Paper, New Delhi.

——— (1991). *Industry and Environment: Select Issues*, New Delhi.

Ast, J.A. (1993). *Milieukunde en milieubeleid* (Environmental science and policy), Samson H.D. Tjeenk Willink, Alphen aan de Rijn.

Banerjee, S.N. (1991). 'Pollution Control in Cement Industry', in D.H. pai Panandikar ed. *Pollution Control in Indian Industry*, BR Publishers, Delhi.

Baumol, W.J. and W.E. Oates (1971). The use of standards and prices for the protection of the environment, *Swedish Journal of Economics*, 73(1), pp. 42–54.

——— (1990). *The Theory of Environmental Policy*, Cambridge University Press, Cambridge.

Berg, K. van den (1994). The possibility of using economic instruments in the fertiliser industry in India, Master's thesis in environmental economics, Vrije Universiteit, Amsterdam.

Bhat, K.N. (1992). Bhopal Disaster—A Review, *Tijdschrift voor Milieu Aansprakelijkheid*, February, pp. 41–45.

Boersema, J.J., J.W. Copius Peereboom and W.T. de Groot eds (1991). *Basisboek milieukunde* (Textbook in environmental science), Boom, Meppel.

Bohm, P. and C.S. Russell (1985). 'Alternative Policy Instruments', in A.V. Kneese et al. eds., *Handbook of Natural Resource and Energy Economics*, Vol. I, North-Holland, Amsterdam.

Bovenberg, A.L., M.E.T. van den Broek and R.J. Mulder (1991). *Instrumentkeuze in het milieubeleid* (Instrument choice in environmental policy), Ministerie van Economische Zaken, Discussienota 9102, The Hague.

Bowonder, B. and S.S. Arvind (1989). Environmental regulations and litigation in India, *Project Appraisal*, 4(4), pp. 182–96.

Bressers, J.Th.A. (1983). *Beleidseffectiviteit en waterkwaliteitsbeleid* (Policy effectiveness and water quality policy), Technical University Twente, Enschede.

——— (1988). A comparison of the effectiveness of incentives and directives: The case of the Dutch water policy, *Policy Study Review*, 7, pp. 500–518.

——— (1991). Vernieuwingsstrategieën inzake milieubeleid (Strategies of innovation in environmental policy), *Openbaar bestuur*, 1991, No. 5, p. 13.

Brussaard, W., Th.G. Drupsteen, P.C. Gilhuis and N.S.J. Koeman eds (1993). *Milieurecht* (Environmental legislation), W.E.J. Tjeenk Willink, Zwolle.

Buchanan, J.M. and G. Tullock (1975). Polluters' Profits and Political Response: Direct Controls versus Taxes, in *American Economic Review*, 65(1), pp. 139–47.

Cairncross, F. (1991). *Costing the Earth*, The Economist Books, London.

CBS (1992), *Statistisch jaarboek 1992* (Statistical yearbook 1992), Central Bureau of Statistics, Voorburg.

——— (1993a). Vijf milieuwetten. Vergunningverlening en handhaving door de provincies, 1991 (Five environmental Acts. Permit granting and enforcement by the provinces, 1991), *Kwartaalbericht milieustatistieken* (Quarterly report on environmental statistics) 1993/4, pp. 30–38, Central Bureau of Statistics, Voorburg.

——— (1993b). Wet verontreiniging oppervlaktewateren (WVO). Vergunningverlening en handhaving door de waterkwaliteitsbeheerders, 1990–1991 (Surface Water Pollution Act. Permit granting and enforcement by water quality boards, 1990–1991). *Kwartaalbericht milieustatistieken* (Quarterly report on environ- mental statistics) 1993/3, pp. 31–37. Central Bureau of Statistics, Voorburg.

——— (1993c). *Milieukosten van bedrijven 1991* (Environmental costs of firms, 1991), Central Bureau of Statistics, Voorburg.

——— (1993d). *Statistisch jaarboek 1993* (Statistical yearbook 1993), Central Bureau of Statistics, Voorburg.

——— (1993e). *Samenvattend overzicht van de industrie 1991* (Industrial survey 1991), Central Bureau of Statistics, Voorburg.

——— (1994a). *Statistisch jaarboek 1994* (Statistical yearbook 1994), Central Bureau of Statistics, Voorburg.

——— (1994b). Kosten en financiering van het milieubeheer 1991 (Costs and financing of environmental management 1991), *Kwartaalbericht Milieustatistieken* (Quarterly report on environmental statistics) 1994/2, Central Bureau of Statistics, Voorburg.

——— (1994c). *Milieustatistieken voor Nederland 1994* (Environmental Statistics for the Netherlands 1994), Central Bureau of Statistics, Voorburg.

Central Pollution Control Board (CPCB) (1992). *Pollution Control Acts, Rules and Notifications Issued Thereunder*, New Delhi.

Chandramouli, K.V. et al. (1993). 'Some Little Known and Forgotten Dye Techniques in Indian Textiles', *Congress on Traditional Sciences and Technologies: Proceedings*, 28 November–3 December, Indian Institute of Technology, Bombay, Tex-1.

Chopra, Kanchan, Gopal K. Kadekodi and Nandita Mongia (1993). Water Pollution Abatement Costs—A Cross Industry Analysis, in the same authors'

Environmental Impacts of Projects: Planning and Policy Issues, Institute of Economic Growth, Delhi, Working Paper E/158/93, pp. 85–94.

CPB (1992). Long Term Economic Consequences of Energy Charges (in Dutch). CPB Working Documents No. 43, The Hague.

Dales, J.H. (1968). *Pollution, Property and Prices*, University of Toronto Press, Toronto.

Downing, P.B. (1984). *Environmental Economics and Policy*, Little, Brown and Company, Boston.

Dudek, D.J. and J. Palmisano (1988). Emissions Trading: Why is this Thoroughbred Hobbled? *Columbia Journal of Environmental Literature*, 217.

Federation of Indian Chambers of Commerce and Industry (FICCI) (1989). National Conference on Pollution Control, 3 January, New Delhi.

Glasbergen, P. ed. (1989). *Milieubeleid: theorie en praktijk* (Environmental policy, theory and practice), VUGA, The Hague.

Government of India (1980). *Report of the Committee for Recommending Legislative Measures and Administrative Machinery for Ensuring Environmental Protection* (Chairman: N.D. Tiwari), Department of Science and Technology, 15 September, New Delhi.

——— (1988). *Fiscal Incentives for Environmental Protection*, Department of Environment, New Delhi.

——— (February 1992). *Policy Statement for Abatement of Pollution*, Ministry of Environment and Forests, New Delhi.

——— (June 1992). *National Conservation Strategy and Policy Statement on Environment and Development*, Ministry of Environment and Forests, New Delhi.

Hahn, R.W. (1989). Economic Prescriptions for Environmental Problems: How the Patient Followed the Doctor's Orders, *Journal of Economic Perspectives*, 3(2), pp. 95–114.

Hanley, N., S. Hallett and I. Moffatt (1990). Why is More Notice Not Taken of Economists' Prescriptions for the Control of Pollution? *Environment and Planning A*, 22, pp. 1421–39.

Houtsma, W.H. (1994). Het herenakkoord als reflex (The gentleman's agreement as a reflex), in *Milieudefensie*, 1994, No. 2, pp. 6–9.

Indian Institute of Environment and Ecology (IIEE) (1991). *The Indian Directory of Environment*, New Delhi.

Jansen, K. (1991). Structural Adjustment and Sustainable Development, *Internationale Spectator* 45(11), pp. 696–701.

Janssen, L.H.J.M., P.A. Okken and J. van Ham (1991). *Praktijk van de CO_2 emissiebeperking* (Practice of CO_2 emission reduction), Kluwer, Deventer.

Joshi, H.N. and T.S. Chouhan (1991). *Environmental Monitoring and Pollution Measures*, Academic Publishers, Jaipur.

Karnataka State Pollution Control Board (KSPCB) (1993). *Annual Report 1992–93*, Bangalore.

——— (1994). *Annual Report 1993–94*, Bangalore.

Klok, P.J. (1989). *Convenanten als instrument van milieubeleid* (Covenants as an instrument of environmental policy), Faculteit der Bestuurskunde, Universiteit Twente, Enschede.

Klok, P.J., J. Ligteringen, H. Pullen and J. Schuddeboom (1992). *Financiële instrumenten in het beleidsproces* (Financial instruments in the policy process), VROM/ECW advice No. 5, The Hague.

Kothari, Miloon and Ashish Kothari (1992). 'Heading for Trouble—Manmohanomics and the Environment', *Frontline*, 28 August, pp. 78–81.

Kothari's Industrial Directory of India. 37th edn. (1992). Kothari, Madras.

Kothari's Yearbook on Business and Industry 1991 (1991). Kothari, Madras.

KPMG/IVA (1993). *NMP evaluatie industrie* (Evaluation of the NEPP for the industry), KPMG Milieu/IVA-Tilburg, The Hague.

Lal's (1990). *Lal's Commentaries on Water and Air Pollution Laws along with Environment Protection Act and Rules, 1986*, revised by Gyanandra Kumar, K.B. Asthana and S.K. Gadi, 3rd edition, Law Publishers India, Allahabad.

Langeweg, F. ed. (1988). *Concern for Tomorrow—National Environmental Survey 1985–2010*, Samson H.D. Tjeenk Willink, Alphen aan de Rijn.

Milieuvoorschriften (1981–93). *Milieuvoorschriften* (Environmental regulation), Delwel, The Hague.

Mukherjee, Amitava (1993). *Structural Adjustment Programme: Putting the First Things, Poverty and Environment, Last*, Segment Books, New Delhi.

OECD (1986). *OECD and the Environment*, OECD, Paris.

——— (1990). OECD economic surveys. *The Netherlands 1989/1990*, OECD, Paris.

——— (1991). *Environmental Policy: How to Apply Economic Instruments*, OECD, Paris.

——— (1993a). OECD economic surveys. *The Netherlands 1992/1993*, OECD, Paris.

——— (1993b). *Pollution Abatement and Control Expenditures in OECD Countries*. Environment Monographs, No. 75. OECD, Paris.

——— (1995). *Environmental Performance Reviews: Netherlands*, OECD, Paris.

Opschoor, J.B. and J.B. Vos (1989). *Economic Instruments for Environmental Protection*, OECD, Paris.

Opschoor, J.B., A.F. de Savornin Lohman and H.B. Vos (1994). *Managing the Environment. The Role of Economic Instruments*, OECD, Paris.

Opschoor, J.B. and R.K. Turner eds (1994). *Economic Incentives and Environmental Policies: Principles and Practice*, Kluwer Academic Publishers, Dordrecht.

Pigou, A.C. (first published in 1920). *The Economics of Welfare*, Macmillan, London.

Ravichandran, M. (1989). Economics of Pollution Prevention and Control: A Study of Industrial Pollution in Karnataka, Ph.D. Thesis submitted to the University of Mysore, Mysore (unpublished) (prepared at the Institute for Social and Economic Change, Bangalore).

RIVM (1988). *Evaluatierapport verzuring* (Evaluation report on acidification).

——— (1991). *Nationale Milieuverkenning 2: 1990–2010* (National Environmental Survey). Samsom H.D. Tjeenk Willink, Alphen aan de Rijn.

——— (1993). *Nationale Milieuverkenning 3: 1993–2015* (National Environmental Survey). Samsom H.D. Tjeenk Willink, Alphen aan de Rijn.

Rosencranz, Armin, Shyam Divan and Martha L. Noble (1991). *Environmental Law and Policy in India: Cases, Materials and Statutes*, Tripathi, Bombay.

Royston, Michael G. (1979). *Pollution Prevention Pays*, Pergamon Press, Oxford.

Savornin Lohman, A.F. de (1994). *Workshop on implementation of environmental taxes: Case study on the Netherlands*, OECD, Paris.

Schuurman, J. (1988). *De prijs van water* (The price of water), Gouda Quint B.V., Arnhem.

Shah, Rakesh (1992). Pollution Control Equipment, Availability, Uses and Misuses, *Economic Times*, 6 June.

Shane, N. Jeffrey (1979). Environmental Law in the Developing Nations of Southeast Asia, in Colin MacAndrews and Chia Lin Sien eds. *Developing Economics and the Environment, The South Asian Experience*, McGraw-Hill Southeast Asia Series, Singapore.

Stavins, R.N. and B.W. Whitehead (1992). Dealing with Pollution. Market-Based Incentives for Environmental Protection, *Environment*, 34(7), pp. 7–11 and 29–42.

Storm, S.T.H. and C.W.M. Naastepad (1991). Beleidsopties voor India (Policy Options for India), *Economisch Statistische Berichten*, 9 October 1991, pp. 1009–13.

The Hindu Survey of the Environment 1993. The Hindu, Madras.

Tietenberg, T.H. (1985). *Emissions Trading: An Exercise in Reforming Pollution Policy*, Resources for the Future, Washington, D.C.

——— (1990). Economic Instruments for Environmental Regulation, *Oxford Review of Economic Policy*, Vol. 6(1), Spring, pp. 17–33.

——— (1992a). *Environmental and Resource Economics*, HarperCollins Publishers, New York.

——— ed. (1992b). *Innovation in Environmental Policy*, Edward Elgar Publishing Ltd., Hants.

Twijnstra Gudde (1982). Verkenning van de kosten en effecten van de vergunningsverlening in het Rijnmondgebied (Exploring costs and effects of permits in the Rijnmond area), a report, Deventer.

Venugopal, B.R. (1993a). Hazards of Synthetic Dyes and the Potential for Natural Dyes, *Congress on Traditional Sciences and Technologies, Proceedings*, 28 November–3 December, Indian Institute of Technology, Bombay, Tex-17.

——— (1993b). Use of Nontoxic and Least Pollutant Dyes in India, *Colourage*, Tiruppur Seminar cum Exhibition Special, November 1993, pp. 65–66.

Verbruggen, H. (1991). Political Economy Aspects of Environmental Policy Instruments, in F. Dietz, F. van der Ploeg and J. van der Straaten eds, *Environmental Policy and the Economy*, Elsevier Science Publishers, Amsterdam.

——— (1993). The Trade Effects of Economic Instruments, in OECD, *Environmental Policies and Industrial Competitiveness*, OECD, Paris.

VoMil (1972). *Urgentienota milieuhygiëne* (Emergency memorandum environmental hygiene), Second Chamber 1971–1972, 11906, Nos. 1–2, SDU, The Hague.

——— (1974). *Nota Instrumenten milieuhygiënisch beleid, heffingen en fysieke reguleringen* (Report on instruments of environmental policy, charges and direct regulation), Second Chamber 1974–1975, No. 13100. SDU, The Hague.

VROM (1983). *Plan Integratie Milieubeleid* (Plan integration environmental policy). Second Chamber 1982–1983, No. 17931–6. SDU, The Hague.

——— (1985a). *Indicatief Meerjaren Programma Milieubeheer 1986–1990* (Indicative long-range Program Environmental Management 1986–1990), Second Chamber 1985–1986, 19204, Nos. 1–2, SDU, The Hague.

VROM (1985b). *Indicatief Meerjaren Programma Milieubeheer 1986–1990—Voortgangsrapportage* (Indicative long-range Program Environmental Management 1986–1990 Progress Report), Second Chamber 1985–1986, 19204, No. 3, SDU, The Hague.

——— (1987). *Milieuprogramma 1988–1991, Voortgangsrapportage* (Environmental program 1988–1991, Progress report), Second Chamber 1987–1988, 20202, Nos. 1–2, SDU, The Hague.

——— (1989). *National Environmental Policy Plan; To Choose or Lose*, Second Chamber 1988–1989, 21137, Nos. 1–2. SDU, The Hague.

——— (1990). *National Environmental Policy Plan–Plus*, Second Chamber 1989–1990. 21137, Nos. 20–21. SDU, The Hague.

——— (1993). *National Environmental Policy Plan 2*, Second Chamber 1993–1994, SDU, The Hague.

——— (1996). *Milieuprogramma 1997–2000* (Environmental programme 1997–2000), Second Chamber 1996–1997, 25005, Nos. 1–2, SDU, The Hague.

Wiersma, D. (1989). De efficiëntie van een marktconform milieubeleid. Een uitwerking van de SO_2 emissiebestrijding van de Nederlandse elektriciteitssector (The efficiency of a market-oriented environmental policy. An elaboration of the SO_2 emission abatement in the Dutch electricity sector), Dissertation, State University Groningen.

World Bank (1992). *World Development Report 1992*, Washington, D.C.

World Commission on Environment and Development (WCED) (1987). *Our Common Future*, Oxford University Press, Oxford.

WRR (1992). *Milieubeleid: Strategie, instrumenten en handhaafbaarheid* (Environmental policy: Strategy, instruments and enforceability), SDU, The Hague.

Zwerver, S., M. Bovenkerk and P.J. Mak (1984). Verzuring: nationale en internationale ontwikkelingen (Acidification: National and international developments), in E.H. Adema en J. van Ham eds: *Zure regen, oorzaken, effecten en beleid* (Acid rain, causes, results and policy), Pudoc, Wageningen.

INDEX

abatement costs, 30, 36, 55, 56
Abubacker, M.H., 182
accelerated depreciation *see* depreciation allowances
acceptability, 33
acidification, 24, 123, 147ff
add-on technologies *see* 'end-of-pipe' technologies
administrative costs, 33
Adriaanse, A., 146ff
Agenda 21, 77
agriculture, 23, 28, 67, 70, 78, 116–17, 149, 150, 157
air pollution, 71, 74, 119, 133–34
'alara' principle, 131, 151
Amelrooy, M. van, 66, 172ff
Arentsen, M.J., 129, 133, 134, 137
Arunachalam, M., 113
Arvind, S.S., 29

Banerjee, S.N., 183
Bangarappa, S., 113
bans, 18, 129, 130, 134
Baumol, W.J., 29, 34, 37, 39, 57
Berg, K. van den, 201
Best Available Technology, 41, 76, 102, 124, 201, 226, 232
best practicable means, 131, 134, 226
best technical means, 131, 134, 226
Bhat, K.N., 29
Bhopal, 74, 76, 96
biodiversity, 25, 77
Boersema, J.J., 138
Bohm, P., 22, 34, 35, 37
Bovenberg, A.L., 34, 35, 42ff, 57, 195, 201
Bowonder, B., 29
Bressers, J. Th.A., 54, 132, 140
Brundtland Commission *see* World Commission on Environment and Development
Brussaard, W., 131, 132
Buchanan, J.M., 52

Cairncross, F., 58
carbon dioxide, 104
cash crops, 28
cement industry, 71, 90, 108–9, 168–71, 180ff
Chandramouli, K.V., 180
charges, 18, 21, 24, 31, 36, 46–49, 53–55, 102, 114, 138, 147, 151, 165, 194, 195, 201, 232
Chopra, K., 202
citizens' suit, 87–88, 94
'clean' technologies, 83, 95, 97, 104–5, 121, 161
climate change, 24, 77, 123, 144ff, 236
Club of Rome, 118
'command and control', 22, 24, 26, 30, 40, 50–52, 54–55
common/collective treatment plants, 96, 103, 113, 231
communicative instruments, 17, 24, 31, 34, 50, 135ff, 166, 192, 201
compliance (costs), 98, 104ff, 137, 160ff, 228–29
consent, 80–81
consumers, 23
cost effectiveness, 20, 22
covenants *see* voluntary agreements
'cradle-to-grave' approach, 24

Dales, J.H., 57
damage compensation, 27
Declaration of Rio, 21
deforestation, 25, 77
dehydration, 123, 155–56
deposit–refund systems, 18, 21, 39, 46, 48, 165
depreciation allowances, 26, 90, 143
direct regulation, 18, 31, 46–50, 52–55, 77, 129ff, 165, 190ff, 236
'double dividend', 29
Downing, P.B., 34, 35, 56
Dudek, D.J., 38

economic instruments, passim
economic reforms, 112ff
education, 135
effect-oriented policy, 120ff
effectiveness (environmental -), 31
efficiency, 20, 22, 31, 32
effluent treatment (plant), 173ff
'end-of-pipe' technologies, 83, 95, 97, 104–5, 121, 161
energy, 24, 70
enforcement (costs), 23, 30, 31, 33, 49, 51, 70, 85–87, 104ff, 119, 159–60, 193, 227–29
environmental audit, 76, 95
environmental impact assessment, 69, 72, 100, 130, 227
environmental management (systems), 18, 137–38, 153
equity, 32
European Communities *see* European Union
European Union, 21, 28, 39, 128
eutrophication, 123, 150–51
external effects, 29

fertiliser (industry), 27, 90, 108–9, 168–71, 186ff
fisheries, 78
forestry, 70, 77

Gandhi, I., 68, 72
Ganga Action Plan, 85
General Agreement on Tariffs and Trade (GATT), 28
Glasbergen, P., 131, 133
global warming *see* climate change
greenhouse effect *see* climate change
green revolution, 186

Hahn, R.W., 29, 55
Hanley, N., 52
hazardous/harmful substances/waste, 76, 91–92, 93, 96, 100, 123, 151–52
health, 81, 155
households, 25
Houtsma, W.H., 136–37
hydropower, 25

income distribution, 32
industrialisation, 25, 116
industry, 25, 28, 67, 78, 146, 149, 157
information provision, 18
innovation, 23, 36, 239
'integrated chain management', 24, 153
integrated pollution control, 93, 121ff
International Monetary Fund (IMF), 27
investment allowance, 91

Jansen, K., 29
Janssen, L.H.J.M., 147
Joshi, H.N., 182
judicial instruments, 31

Klok, P.J., 136, 137, 143
Kothari, A., 28, 112
Kothari, M., 28, 112

land degradation, 25
Langeweg, F., 124
liability, 18, 24, 40, 48, 49, 56, 91
liberalisation, 27, 28, 29, 67, 112ff

marketable permits *see* tradable permits,
market based approaches/instruments, 25, 31
market distortions/imperfections, 40, 47
maximum acceptable risk, 151, 155

means-prescriptions, 130
Mehta, M.C., 26
mineral extraction and processing, 70
mining, 25
Ministry of Environment and Forests (India), 75ff
Ministry of Housing, Physical Planning and Environment (Netherlands), 127ff
mobile sources, 23
monitoring (costs), 23, 55, 56, 81, 84, 88, 100–1, 119, 189, 230–31
Montreal Protocol, 29, 77, 147
moral suasion, 18, 35
morbidity, 71
Mukherjee, A., 112

Naastepad, C.W.M., 27
National Environmental Policy Plan (Netherlands), 122ff
natural resources, 24, 123, 156
No Objection Certificate, 80, 92
noise, 76, 123, 154–55
non-governmental organisations (NGOs), 26, 56, 78, 94, 118, 226
non-point sources, 23

Oates, W.E., 29, 34, 37, 39, 57
Opschoor, J.B., 17, 22, 35, 36, 52, 54, 140, 166
Organisation for Economic Co-operation and Development (OECD), 21, 31, 33, 39
'Our Common Future', 20, 25
overfishing, 25
overgrazing, 25
ozone depletion (stratospheric -), 24, 77, 123, 147

Palmisano, J., 38
permits, 119, 129ff
pesticides, 74, 103, 152
Pigou, A.C., 22, 57
point-sources, 23
polluter pays principle, 19, 20, 21, 26, 28, 33, 34, 39, 52, 71, 76, 119, 124, 143, 226, 235
Pollution Control Boards, 73ff
pollution control/treatment equipment/measures/plants, 81, 96–97, 102, 109
poverty, 68–69
power plants, 23, 25, 149
precautionary principle, 19, 22, 26
prevention principle, 21, 124
principal agent problem, 40
process-integrated technologies *see* 'clean' technologies
property rights, 27
Public Interest Litigation, 87–88, 94, 98, 226, 231
public procurement, 18

Rathore, K.C., 115
Ravichandran, M., 107–8, 110
raw materials, 24
recycling, 70, 100, 152–53
Reddy, V.S., 115
refineries, 149
regulatory instruments, 40, 47
research and development (R&D), 34, 39, 47, 48, 70, 80
Rio Conference *see* United Nations Conference on Environment and Development
Rosencranz, A., 68–69, 71, 74, 87–88
Royston, M.G., 105
Russell, C.S., 22, 34, 35, 37

salinisation, 25
Savornin Lohman, A.F. de, 142
Schuurman, J., 140
Shah, R., 102
Shane, N.J., 61
Shivalingaiah, B., 115
Singh, M., 27
situation–characteristics, 42ff, 194ff, 240
small-scale industries, 102–3, 113, 193, 231
soft loans, 92
soil contamination, 154
soil erosion, 25
source-oriented approach, 120ff

standards, 18, 41, 46, 48, 49, 53, 74, 75, 79, 82ff, 88, 98, 121, 129, 176, 182, 226–27
stand-still principle, 19, 124
Stavins, R.N., 25
stench nuisance, 155
Stockholm Conference (1972), 68, 72, 224
Storm, S.T.H., 27
structural adjustment, 27–29
subsidies, 18, 21, 26, 33, 39, 46–48, 52, 90–92, 143, 192, 235
Summers, L., 28
Sundareshan, H.R., 115
sustainable development, 29, 69, 122, 125–26, 164–65, 241

tanneries, 88, 96
target-prescriptions, 130
taxes, 18, 21, 27, 53–55, 142, 165
tax differentiation/exemptions/relief, 91, 114, 142–43
textile industry, 90, 108–9, 168ff
Tietenberg, T.H., 30, 31, 49, 50, 164, 233
Tiwari, N.D., 74
tradable (emission) rights/permits, 18, 21, 24, 31, 37, 46, 48, 49, 54, 55, 144, 165, 195–96, 236
traffic, 25, 103, 149
transport, 70, 117, 157
Tullock, G., 52
Turner, R.K., 17, 36, 52, 166

United Nations Conference on Environment and Development (UNCED), 20, 69, 77
United States, 30

Venugopal, B.R., 180
Verbruggen, H., 37, 53
voluntary agreements/commitments, 18, 24, 35, 46, 48, 49, 52, 135ff, 153, 164, 194, 227, 236
Vos, J.B., 22, 54, 140

waste, 25, 123, 152ff
water cess (India), 73, 89–90, 100–1, 228, 234
water pollution, 23–25, 71, 73, 82–85, 98–99, 134
water pollution charge (Netherlands), 139ff, 173ff, 192, 228
water resources, 25, 70
Whitehead, B.W., 25
Wiersma, D., 164
World Bank, 27, 28
World Commission on Environment and Development (WCED), 20, 25, 125

Zwerver, S., 148

The following titles have appeared in the series:

1. H.H. de Haan: *Alternatives in Industrial Development: Sugarcane Processing in India* (1988).
2. Jairus Banaji and Rohini Hensman: *Beyond Multinationalism: Management Policy and Bargaining Relationships in International Companies* (1990).
3. Martin Doornbos, F. van Dorsten, M. Mitra and P. Terhal: *Dairy Aid and Development: India's Operation Flood* (1990).
4. Martin Doornbos and K.N. Nair (eds): *Resources, Institutions and Strategies: Operation Flood and Indian Dairying* (1990).
5. Nirmala Banerjee (ed.): *Indian Women in a Changing Industrial Scenario* (1991).
6. Nirmal Sengupta: *Managing Common Property: Irrigation in India and the Philippines* (1991).
7. Hein Streefkerk and T.K. Moulik (eds): *Managing Rural Development: Health and Energy Programmes in India* (1991).
8. Willem van Schendel: *Three Deltas: Accumulation and Poverty in Rural Burma, Bengal and South India* (1991).
9. Hans Linnemann (ed.): *South-South Trade Preferences: The GSTP and Trade in Manufactures* (1992).
10. T.M. Thomas Issac, P.A. van Stuijvenberg and K.N. Nair: *Modernisation and Employment: The Coir Industry in Kerala* (1992).
11. Kanta Ahuja, Huub Coppens and Herman van der Wusten (eds): *Regime Transformations and Global Realignments: Indo-European Dialogues on the Post-Cold War World* (1993).
12. Pitou van Dijck and K.S. Chalapati Rao: *India's Trade Policy and the Export Performance of Industry* (1994).
13. G.K. Chadha: *Employment, Earnings and Poverty: A Study of Rural India and Indonesia* (1994).
14. Indira Hirway and Piet Terhal: *Towards Employment Guarantee in India: Indian and International Experiences in Rural Public Works Programmes* (1994).
15. V.R. Panchamukhi et al.: *Complementarity in Trade and Production: Intra-South Potentials* (1995).
16. Loes Schenk-Sandbergen: *Women and Seasonal Labour Migration* (1995).
17. C.H. Hanumantha Rao and Hans Linnemann: *Economic Reforms and Poverty Alleviation in India* (1996).
18. G.K. Lieten: *Development, Devolution and Democracy: Village Discourse in West Bengal* (1996).
19. Martin Doornbos and Sudipta Kaviraj (eds): *Dynamics of State Formation: India and Europe Compared* (1997).
20. Atul Sarma, Gerrit Faber and Pradeep Kumar Mehta: *Meeting the Challenges of the European Union: Prospects of Indian Exports* (1997).